CHEMISTRY RESEARCH AND APPLICATIONS

FURAN

CHEMISTRY, SYNTHESIS AND SAFETY

CHEMISTRY RESEARCH AND APPLICATIONS

Additional books and e-books in this series can be found on Nova's website under the Series tab.

CHEMISTRY RESEARCH AND APPLICATIONS

FURAN

CHEMISTRY, SYNTHESIS AND SAFETY

IDA BAILEY
EDITOR

Copyright © 2019 by Nova Science Publishers, Inc.

All rights reserved. No part of this book may be reproduced, stored in a retrieval system or transmitted in any form or by any means: electronic, electrostatic, magnetic, tape, mechanical photocopying, recording or otherwise without the written permission of the Publisher.

We have partnered with Copyright Clearance Center to make it easy for you to obtain permissions to reuse content from this publication. Simply navigate to this publication's page on Nova's website and locate the "Get Permission" button below the title description. This button is linked directly to the title's permission page on copyright.com. Alternatively, you can visit copyright.com and search by title, ISBN, or ISSN.

For further questions about using the service on copyright.com, please contact:
Copyright Clearance Center
Phone: +1-(978) 750-8400   Fax: +1-(978) 750-4470   E-mail: info@copyright.com.

NOTICE TO THE READER

The Publisher has taken reasonable care in the preparation of this book, but makes no expressed or implied warranty of any kind and assumes no responsibility for any errors or omissions. No liability is assumed for incidental or consequential damages in connection with or arising out of information contained in this book. The Publisher shall not be liable for any special, consequential, or exemplary damages resulting, in whole or in part, from the readers' use of, or reliance upon, this material. Any parts of this book based on government reports are so indicated and copyright is claimed for those parts to the extent applicable to compilations of such works.

Independent verification should be sought for any data, advice or recommendations contained in this book. In addition, no responsibility is assumed by the Publisher for any injury and/or damage to persons or property arising from any methods, products, instructions, ideas or otherwise contained in this publication.

This publication is designed to provide accurate and authoritative information with regard to the subject matter covered herein. It is sold with the clear understanding that the Publisher is not engaged in rendering legal or any other professional services. If legal or any other expert assistance is required, the services of a competent person should be sought. FROM A DECLARATION OF PARTICIPANTS JOINTLY ADOPTED BY A COMMITTEE OF THE AMERICAN BAR ASSOCIATION AND A COMMITTEE OF PUBLISHERS.

Additional color graphics may be available in the e-book version of this book.

Library of Congress Cataloging-in-Publication Data

ISBN: 978-1-53615-390-3

Published by Nova Science Publishers, Inc. † New York

Contents

Preface vii

Chapter 1 Homogeneous Catalytic Reactions of Furans and
Formylfurans with Aqueous Hydrogen Peroxide **1**
*Larisa A. Badovskaya, Vladimir V. Poskonin
and Larisa V. Povarova*

Chapter 2 Effective Synthesis of Novel Furan-Fused
Pentacyclic Triterpenoids by Base-Promoted or
Gold-Catalyzed 5-Exo Dig Heterocyclization of
2-Alkynyl-3-Oxotriterpene Acids **67**
A. Yu. Spivak and R. R. Gubaidullin

Chapter 3 Titanium Catalysis in the Chemistry of Furanes **91**
*Leila O. Khafizova, Mariya G. Shaibakova
and Usein M. Dzhemilev*

Index **127**

Related Nova Publications **133**

PREFACE

In this compilation, the authors discuss the reactions of furans and formylfurans with aqueous hydrogen peroxide which are caused by peculiarities in the chemical behavior of these reagents, as well as their availability. It is known that the furan cycle can demonstrate properties of aromatic compounds, dienes and enoloethers depending on their structure and reaction conditions.

Furans are used in medicinal chemistry as useful intermediates in the synthetic transformations aimed at the development of new pharmaceutical agents. As such, the authors also disclose the direct and atom-economical synthetic route to new [3,2-b]furan-fused pentacyclic triterpenoids via 5-exo-dig cyclization of accessible 2-alkynyl derivatives of betulonic, ursonic and oleanonic acids in the presence of the strong bases or catalyzed by transition gold complexes.

The closing chapter presents research data published by the authors in the last five years on the new developed one-pot synthesis of tetrasubstituted furans by the reaction of symmetrical and unsymmetrical acetylenes with carboxylic acid esters and $EtAlCl_2$ in the presence of metallic magnesium and the Cp_2TiCl_2 catalyst.

Chapter 1 - The reactions of furans and formylfurans with aqueous hydrogen peroxide are attracted by the peculiarities of the chemical behavior of these reagents, as well as their availability. It is known that the furan cycle in the molecules of furans can prove properties of aromatic compounds, dienes and enoloethers depending on their structure and the reaction conditions. In turn, hydrogen peroxide during its decomposition can form various types of reaction particles, which participate in the oxidation processes. These facts, combined with the unique properties of the water involved in the above reactions, make it possible to control their direction by changing the reaction factors.

Chapter 2 - The available plant metabolites, that are betulinic, ursolic and oleanolic acids, represent an important class of biologically active substances, which are in high demand in medicinal chemistry. These compounds are of interest for pharmacological investigations, as they exhibit various activities, such as anti-inflammatory, antiviral, hepatoprotective, antiparasitic and anticancer effects. Owing to the presence of easily transformable functional groups (OH, COOH, C=C), pentacyclic triterpenoids possess a good synthetic potential and are actively used as promising structural models for the discovery of new drugs. Currently, in order to increase the biological potential and bioavailability of native triterpene acids, their numerous synthetic analogues have been prepared. Among them, considerable attention has been given to heterocyclic triterpenoids with various nitrogen, sulfur, and oxygen heterocycles once they have been studied as antitumor, antiosteoporosis, anti-inflammatory and antileishmanial agents. Among this group of compounds, furan triterpenoid derivatives have not been reported in the literature. Meanwhile, polysubstituted furans represent an important class of oxygen heterocycles and occur as structural moieties in many natural products and pharmaceutically important substances. Furans are used in medicinal chemistry as useful intermediates in the synthetic transformations aimed at the development of new pharmaceutical agents. This review discloses the direct and atom-economical synthetic route to new [3,2-*b*]furan-fused pentacyclic triterpenoids via 5-exo-dig cyclization of accessible 2-alkynyl derivatives of betulonic, ursonic and oleanonic acids in the presence of the strong bases or catalyzed by transition gold complexes. Good prospects in these reactions have been revealed for highly alkynophilic gold-containing catalytic system $PPh_3AuCl/AgOTf$.

Triterpenoids with a terminal or internal triple bond undergone cycloisomerization in the presence of the $PPh_3AuCl/AgOTf$ catalyst under very mild conditions. The generality of the method was demonstrated by the efficient preparation of furanfused triterpenoids, containing various functional groups, including hard Lewis base-sensitive groups such as CN, NO_2, or OAc, at the C-5' atom of the furan ring.

Chapter 3 - This chapter presents the research data published by the authors in the last five years on the new developed one-pot synthesis of tetrasubstituted furans by the reaction of symmetrical and unsymmetrical acetylenes with carboxylic acid esters and $EtAlCl_2$ in the presence of metallic magnesium (acceptor of chloride ions) and the Cp_2TiCl_2 catalyst. The review considers the effect of the structure of initial acetylenes and carboxylic acid esters on the reaction direction and the yield of tetrasubstituted furans. The reaction of symmetric acetylenes with esters of α,ω-dicarboxylic acids and $EtAlCl_2$ is described under the action of the Cp_2TiCl_2 catalyst. It was shown that the length of the hydrocarbon chain between the carboxyl groups in α,ω-dicarboxylic acid esters has a specific effect on the chemoselectivity of the reaction. The Cp_2TiCl_2-catalyzed reaction of symmetrical acetylenes with α,ω-dicarboxylic acid esters in the presence of $EtAlCl_2$ can produce C_5-C_6 cyclic ketones containing alkylidene and alkenyl substituents.

In: Furan: Chemistry, Synthesis and Safety ISBN: 978-1-53615-390-3
Editor: Ida Bailey © 2019 Nova Science Publishers, Inc.

Chapter 1

HOMOGENEOUS CATALYTIC REACTIONS OF FURANS AND FORMYLFURANS WITH AQUEOUS HYDROGEN PEROXIDE

Larisa A. Badovskaya, Vladimir V. Poskonin[*]
and Larisa V. Povarova

Department of Chemistry,
Kuban State Technological University, Krasnodar, Russia

ABSTRACT

The reactions of furans and formylfurans with aqueous hydrogen peroxide are attracted by the peculiarities of the chemical behavior of these reagents, as well as their availability. It is known that the furan cycle in the molecules of furans can prove properties of aromatic compounds, dienes and enoloethers depending on their structure and the reaction conditions. In turn, hydrogen peroxide during its decomposition can form various types of reaction particles, which participate in the oxidation processes. These facts, combined with the unique properties of

---

[*] Corresponding Author's E-mail: vposkonin@mail.ru.

the water involved in the above reactions, make it possible to control their direction by changing the reaction factors.

HOMOGENEOUS REACTIONS OF FURAN AND 2-METHYLFURAN WITH AQUEOUS HYDROGEN PEROXIDE IN THE PRESENCE OF VANADIUM COMPOUNDS

Vapor-phase catalytic oxidation of furans by molecular oxygen in the presence of vanadium catalysts was previously widely studied [1-6]. Maleic anhydride, carbon monoxide and carbon dioxide are the main products of this process, and minor amounts of acetic and oxalic acids are formed. Vapor-phase oxidation of furan occurs more selectively compared with the oxidation of its derivatives: the complete conversion of furan is achieved in a shorter period, and the yield of maleic anhydride is higher.

Photosensitized oxidation of furan and alkylfurans by molecular oxygen in alcoholic media in the presence of various sensitizers (eosin, Bengal rose, methylenblau, etc.) was also studied [7-13]. 5-Alkoxy-2(5H)-furanones, *cis*- and *trans*-isomers of β-formylacrylic acid, 2,5-dialkyl-2,5-dialkoxy-2,5-dihydrofurans, 5-alkoxy- and 5-hydroxy-2(5H)-furanones and other compounds are products of photochemical oxygenation of furan. 2,5-Dialkoxy-2,5-dihydrofurans were previously obtained mainly by the known methods of Clauson-Kaas, based on the reaction of furan alkoxylation in the presence of bromine [14], or on the electrochemical alkoxylation of furan [15-17].

Photosensitized oxidation of furan (1) and its homologues by molecular oxygen under various conditions has been studied in detail [11]. It is experimentally proved that photochemical transformations of furan under the action of molecular oxygen proceed by the following mechanism (Scheme 1.1).

Ozonide 2 was found to mainly form at the initial stage of the process and isolated individually. Product 2 was stable only below -15^{0}C. At slow heating in benzene or petroleum ether, it turned into a more stable dimer under normal conditions. Monosubstituted furan ozonides were found to be less stable and less prone to the formation of dimers in comparison with disubstituted ones. Further transformations of ozonide 2 in the aprotic solvents (CCl_4, $CHCl_3$, $CFCl_3$) are accompanied by a homolytic rupture of the O-O bond and lead to the formation of very unstable intermediates 3 and 4, which pass into di- and monoepoxides 5 and 6.

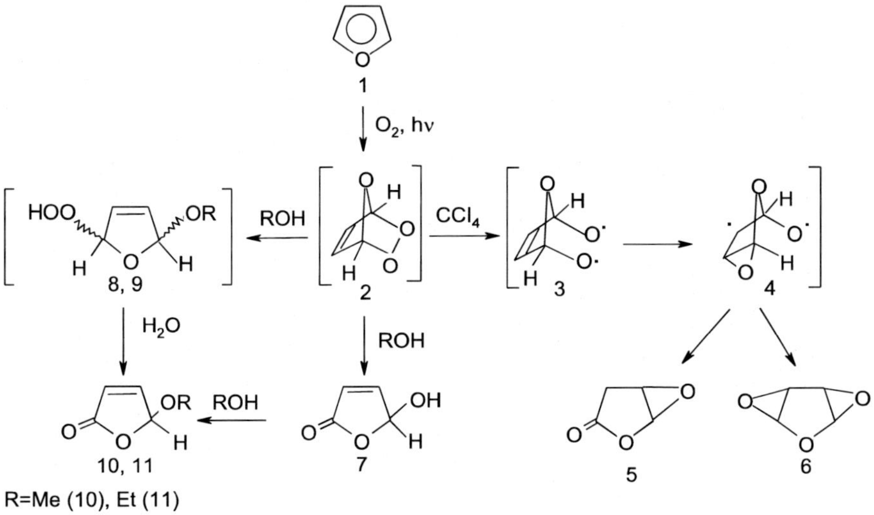

Scheme 1.1.

Ozonide 2 in an alcoholic medium turns into 5-hydroxy-2(5H)-furanone (7a), or into cis- and trans-isomers of 2-alkoxy-5-hydroperoxy-2,5-dihydrofurans 8 and 9, which turn into 5-alkoxy-2(5H)-furanones 10 (5-methoxy) and 11 (5-ethoxy), respectively.

Thus, the intermediate and final products of photosensitized oxidation of furan and its homologues (endoperoxide 2, 2-alkoxy-5-hydroperoxy-2,5-dihydrofurans 8 and 9, 5-alkoxy- and 5-hydroxy-2(5H)-furanones 7a, 10 and 11, diepoxides and epoxyethanes 5 and 6) have been isolated and characterized [11]. Their structure was confirmed by ^{1}H- and ^{13}C-NMR spectroscopy and elemental analysis. The direction of photochemical oxygenation of furan and its homologues was found to be determined by the nature of the solvent, temperature, and structure of the substrate.

The formation and structure of ozonides was confirmed by mass spectrometry of the products of furan, 2-methylfuran and 2,5-dimethylfuran photo-oxidation by molecular oxygen at 20–360^{0}C [12]. The formation of endoperoxides by the 1,4-addition of singlet oxygen was also reported in [13].

The presented data show the processes of vapor-phase and photosensitized oxygenation of furans to pass through the unstable intermediate ozonides formation [11-13].

Synthetic possibilities of vapor-phase oxidation of furan and its homologues by molecular oxygen in the presence of metal-containing catalysts are rather limited [1-6], as they lead to the formation of only maleic anhydride as the main product. In addition, vapor-phase processes are complex and energy-intensive. Sensitized photochemical oxidation of furan compounds [7-13] produces a number of other valuable products, such as mono- and diepoxides, epoxyethane, 5-alkoxy- and 5-hydroxy-2(5H)-furanones, 2,5 dialkoxy-2,5-dihydrofurans, *cis*- and *trans*-isomers of β-formylacrylic acid, and maleic acid. However, photochemical oxidation processes also have a number of significant drawbacks limited their use. This is a long duration of processes, the use of complex special equipment and absolutized solvents, and high energy cost. Therefore, photochemical methods also do not make the above valuable products available.

The oxidation reactions of furan and its homologues by hydrogen peroxide [18-26, 31-35, 39-41] are devoid of a significant part of these disadvantages. These reactions are discussed below. Hydrogen peroxide as an oxidant is very promising due to its high reactivity, availability, and environmental friendliness of the processes occurring with its participation.

The reactions of peroxide oxidation of furan compounds are quite simple in execution. They are the basis for a number of hard-to-reach compounds preparation [19-26, 31-35, 39-41].

Photosensitized oxidation of 2,5-dimethylfuran by hydrogen peroxide in methanol was studied in [19]. The authors suggest the peroxidation of 2,5-dimethylfuran (12) to proceed like its photosensitized oxidation by molecular oxygen [9] with the formation of ozonide 13, which in the presence of methanol is converted into 2,5-dimethyl-2-hydroperoxy-5-methoxy-2,5-dihydrofuran (14) (yield of 74%) [19] (Scheme 1.2).

Reaction of furan with hydrogen peroxide in the presence of sulfuric acid was studied in [20, 21]. Unstable intermediate hydroperoxides were suggested to be intermediates in this reaction. Maleic acid dialdehyde (19), isolated as its 2,4-dinitrophenylhydrazone, was considered as the final product of the reaction. 2-Methylfuran was found to oxidize to the corresponding ketoaldehyde (4-oxopent-2-enal) under these conditions.

Oxidation of furan by hydrogen peroxide in the vapor-gas phase at 160-170°C under pressure was studied in [22]. Under these conditions, a mixture of oxalic, maleic and fumaric acids was obtained. It is assumed that oxygen formed from H_2O_2 acts as an oxidizer in this process.

The authors [23, 24] studied the reactions of furan oxidation with hydrogen peroxide under different conditions. It was shown that organic peroxides were formed at the beginning of the reaction. No heating and the addition of acids required for their formation. At the same time, acid catalysis and heating to 40-100°C required to obtain carbonyl compounds (maleic dialdehyde and monoaldehydes) and carbon acids (maleic, fumaric, and oxalic ones). The increase in the acids concentration in the reaction mixture contributed to the oxidation of carbonyl compounds and the accumulation of dicarboxylic acids. The influence of the solvent nature (water, ethanol, dioxane, tetrahydrofuran, N, N-dimethylformamide, and formic acid) on the direction of the reaction in the furan – H_2O_2 system was revealed. The highest yield of organic acids was observed in the presence of protophilic solvents. The maximum and minimum accumulation of carbonyl compounds occurred in aqueous formic acid and water, respectively.

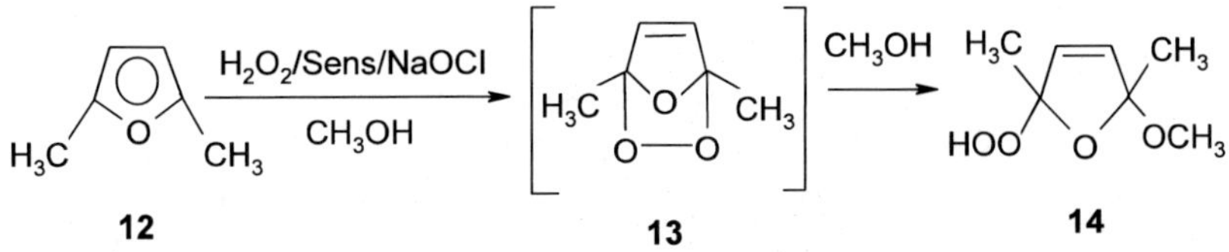

Scheme 1.2.

Thus, the ability to control the direction of the furan reaction with H_2O_2 by changing the reaction factors (temperature, solvent nature, etc.) was first established in [23, 24]. This made it possible to develop new methods for the synthesis of β-formylacrylic, maleic and fumaric acids, based on the oxidation of furan with hydrogen peroxide in the presence of acids and an aqueous organic solvent at 70-75 ^{0}C. However, carbonyl compounds could not be isolated under these conditions. In addition, the oxidation of furan at a temperature above its boiling point was problematic, because it led to the reagent losses and deep oxidative transformations of the target products.

The limited possibilities of this process, due to the insolubility of furan in hydrogen peroxide and its low boiling point, were subsequently overcome. For this purpose, mixed solvents promoting homogenization of the reaction mixture and catalysts leading to the formation of highly reactive particles from hydrogen peroxide were used [25, 26].

Use of vanadium (IV,V) compounds as catalysts was shown [25, 26] to significantly intensify the process of furan oxidation with H_2O_2 and carry it out at lower temperatures. According to [26], catalytic oxidation of furan with hydrogen peroxide is carried out through the stages of radical hydroxylation of the positions 2 and 5 of the furan cycle by OH·radicals formed in the vanadium (IV,V) - H_2O_2 system [27, 28]. This is consistent with the results of studies by EPR methods of the initial stages of furan interaction with hydroxyl radicals generated from the Fenton reagent ($FeSO_4 - H_2O_2$) [29, 30].

The formation of the main reaction products was suggested to pass through the hypothetical 2,5-dihydroxy-2,5-dihydrofuran 18 and the product of its disclosure (dialdehyde 19) formation [26]. The oxidation of furan with aqueous hydrogen peroxide in the presence of compounds of vanadium (IV,V) was shown to form a *cis*-β-formylacrylic acid (in the form of 5-hydroxy-2(5H)-furanone (7a) as its cyclic tautomer), 2(5H)-furanone (16), and maleic 22 and fumaric acids (Scheme 1.3).

The reactions of furan with hydrogen peroxide in aqueous-alcoholic media in the presence of vanadium compounds at temperatures below the boiling point of the organic reagent were studied [31-35].

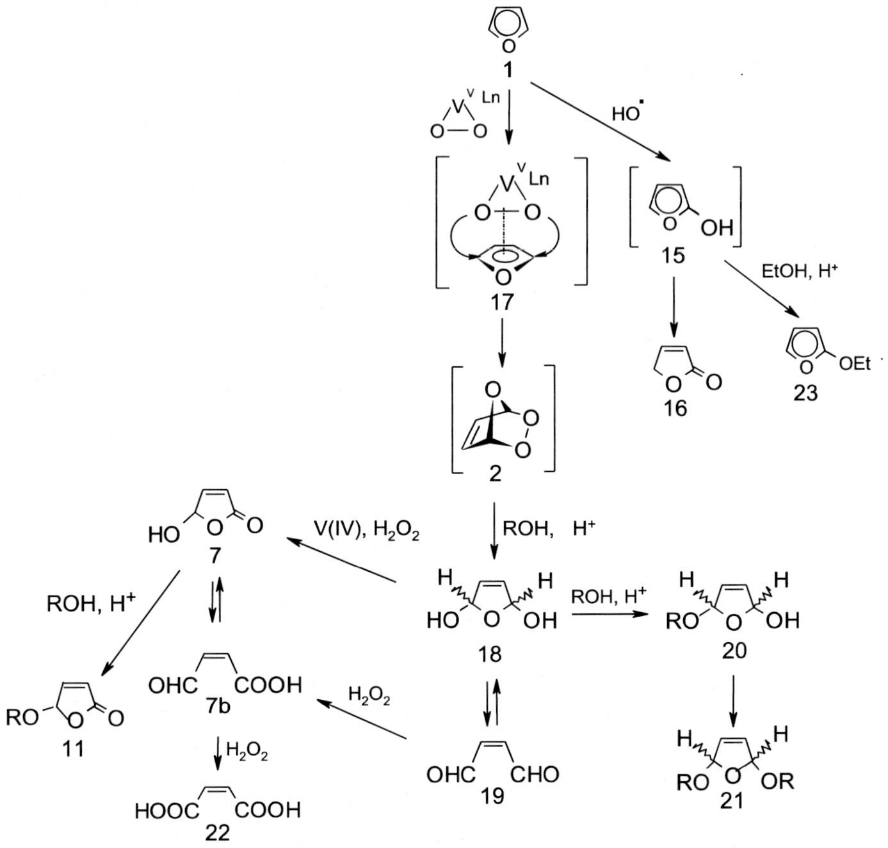

Scheme 1.3.

Furan was oxidized with hydrogen peroxide at 20°C in the presence of vanadium (IV,V) compounds. The reaction mixture was diluted with water or organic solvent (acetone or alcohol) to a concentration of H_2O_2 2 mol/L. Under these conditions, the reaction mixture became homogeneous and had a pH of 5.5, which decreased to 1 during the reaction due to the formation of acids. Oxidation was carried out until the complete conversion of H_2O_2 and peroxides formed (Table 1.1).

In the absence of a vanadium catalyst and organic solvent, the reaction system is heterogeneous and the furan oxidizes very slowly and inefficiently. In this case, an organic peroxide found earlier in [24] is formed in a small amount.

Table 1.1. The influence of solvent nature on the oxidation of furan by hydrogen peroxide in the presence of $VOSO_4$, [furan]: [H_2O_2]: [$VOSO_4$] = 1 : 2 : 0.02, 20°C

Solvent	Total conversion time of H_2O_2, h	Yield, %[b]				
		7	11	21	16	22
H_2O[a]	72.0	–	–	–	–	–
H_2O	22.0	20	–	–	7	10
Acetone	14.0	35	–	–	10	15
Ethanol	6.0	56	6	32	1	–
n-Propanol	7.0	42	6	22	1	–
sec-Propanol	7.5	38	Traces	Traces	2	Traces
n-Butanol	6.5	32	8	20	2	–
tert-Butanol	7.0	56	Traces	Traces	1	Traces
tert-Butanol[c]	24.0	35	–	–	35	–

[a] without $VOSO_4$.
[b] respect to reacted furan.
[c] in the presence of SeO_2, [furan] : [H_2O_2] : [SeO_2] = 1 : 2 : 0.05; 25-30°C.

The use of vanadyl sulfate ($VOSO_4$) as a catalyst reduces the period of complete conversion of H_2O_2 (more than 3 times), but the conversion of furan is only 7%. The reaction products are β-formylacrylic acid (as a mixture of tautomeric forms 7a and 7b), maleic acid (22) and 2(5H)-furanone (16). The overall yield of these compounds is low (Table 1.1).

Furan oxidation is significantly intensified in homogeneous water-organic media (aqueous acetone, aqueous ethanol, aqueous propane-1-ol, or aqueous butane-1-ol). In this case, the period of complete conversion of H_2O_2 is reduced by 3 times, the amount of reacted furan increases by almost 10 times, and the total yield of the reaction products increases significantly [31].

Tautomers of β-formylacrylic acid 7a and 7b, maleic acid (22) and 2(5H)-furanone (16) are the products of furan oxidation in aqueous acetone or aqueous alcohols. In the presence of alcohols the corresponding 5-alkoxy-2(5H)-furanones also formed in a yield of 6-8%. When using propane-2-ol or tert-butanol, these compounds are formed only in trace amounts.

It is characteristic that the yield of tautomers 1 and 2 in aqueous alcohols is increased by 2 times (up to 56%) compared to the furan reaction in aqueous acetone, while acid 3 and furanone 4 are formed in small quantities (yield less than 1%). There is a noticeable accumulation of the corresponding 2,5-dialkoxy-2,5-dihydrofurans 21 as a mixture of *cis*-, *trans*-isomers (*cis*-form significantly predominates) in aqueous alcohols [32-34].

Yields of dialcoxydihydrofurans depend on the alcohol nature and reach the highest value (32%) when using ethanol. When aqueous propane-2-ol or *tert*-butanol are used, these compounds accumulate only in trace amounts (Table 1.1).

It is noteworthy that the yield of β-formylacrylic acid (7) decreases, while the yield of 2(5H)-furanone (16) increases in the presence of SeO_2 [25]. This fact confirms the possibility of controlling the direction of the furan reaction with H_2O_2 by changing the nature of the solvent and catalyst.

The type of vanadium catalyst was found to significantly effect on the yield of products 7, 11, 16, 21a and 22 and the duration of the process (Table 1.2).

The highest yields of the main oxidation products 7, 11, 16, 21a and 22 as well as the shortest reaction time are achieved in the presence of $VOSO_4$ and $VOCl_2$. A noticeable increase in the yield of 2,5-diethoxy-2,5-dihydrofuran (21a) (up to 41%) is observed when using $VOCl_2$. The highest yield of β-formylacrylic acid (7) is achieved in the presence of $VOSO_4$. Yield of 2(5H)-furanone (16) increases in the presence of V_2O_5.

As a result, a method for producing 2,5-dialkoxy-2,5-dihydrofurans by oxidation of furan with hydrogen peroxide in the presence of vanadium compounds in aqueous-alcoholic media has been developed [35]. This new method has advantages over the methods of electrochemical alkoxylation and sensitized photochemical oxygenation of furans [11, 15-17].

The composition of intermediate and final products of furan oxidation with hydrogen peroxide in aqueous ethanol in the presence of $VOSO_4$ was determined by chromatography-mass spectrometry. Formation of the main products 7, 11 and 21a and the intermediates such as 2,5-dihydroxy-2,5-

dihydrofuran (18), 2-hydroxy-5-ethoxy-2,5-dihydrofuran (20) and 2-hydroxyfuran (15) was proved. Compounds 7, 11 and 21a were isolated as their stable derivatives – 2,5-dialkoxy-2,5-dihydrofurans 21 and 2(5H)-furanone (16) [36] (Scheme 1.3).

Table 1.2. The influence of the catalyst nature on the oxidation of furan by hydrogen peroxide in aqueous ethanol, [furan]: [H_2O_2]: [vanadium compound] = 1 : 2 : 0.02, 20°C

Catalyst	Total conversion time of H_2O_2, h	Yield, %[b]				
		7	11	16	21a	22
$VOSO_4$	6	56	6	1	32	–
$VOCl_2$	7	41	9	1	41	traces
$VO(acac)_2$	16	41	8	–	16	traces
V_2O_4	10	33	8	–	16	traces
SeO_2[a]	24	35	–	35	–	–

[a] [фуран] : [H_2O_2] : [SeO_2] = 1 : 2 : 0.05; 25-30°C, water/tert-butanol.
[b] respect to reacted furan.

The authors proposed a scheme of the mechanism of the furan peroxidation in the presence of vanadium compounds [32-34, 38]. The scheme is theoretically justified and confirmed by data on the composition of the final and intermediate reaction products. It provides for the formation of furan endoperoxide 2, 2-hydroxyfuran (15) and 2,5-dihydroxy-2,5-dihydrofuran (18) as the key intermediates [3].

According to the authors [33], furan oxidation takes place with the participation of vanadium peroxocomplexes and hydroxyl radicals in these conditions. These particles are formed by the interaction of hydrogen peroxide with vanadium compounds and have a much greater oxidizing capacity than H_2O_2 [28, 36, 37]. The reaction takes place through the formation of the expected intermediates 2, 15 and 18 in two main directions. These directions provide for the attack of the furan nucleus by OH· radicals and peroxovanadium complexes, respectively (Scheme 1.4). Formation of 2-hydroxyfuran (15) can be explained by the interaction of furan with hydroxyl radicals. The intermediate 15 is converted to a more stable 2(5H)-furanone (16).

The second direction involves the interaction of furan with peroxoform of vanadium catalyst. The reaction system V^{+5}-H_2O_2 at pH < 7 is known to generate a singlet oxygen in the form of a complex V^{+5} ($'O_2$) 17 [36-38] The interaction of this complex with the furan cycle leads to the addition of singlet oxygen to form ozonide 2, which is converted into 2,5-dihydroxy-2,5-dihydrofuran (18). Acid-catalyzed alcoholysis of intermediate 18 leads to 2-hydroxy-5-ethoxy-2,5-dihydrofuran (20), and then to 2,5-diethoxy-2,5-dihydrofuran (21a) as a mixture of *cis*- and *trans*-isomers (*cis*-form prevails) [37].

Dihydroxydihydrofuran 18 hydrolyses to maleic dialdehyde (19). The latter is partially oxidized to β-formylacrylic acid (7b), which is converted to maleic acid (22) in the interaction with the hydrogen peroxide.

Unstable hydroxyfuran 15 was isolated as 2(5H)-furanone (16) as its more stable tautomer. The method of chromatography-mass spectrometry allowed us to detect 2-ethoxyfuran (23) in a mixture of furan oxidation products in aqueous ethanol. This compound, apparently, is the product of hydroxyfuran 15 alkoxylation.

The results of studies of 2-methylfuran oxidation by hydrogen peroxide in the system of aqueous alcohol – vanadium compound have been presented in [38-40].

This reaction was carried out under conditions similar to furan oxidation (20-25°C, molar ratio 2-methylfuran, H_2O_2 and catalyst = 1 : 2 : 0.02). The pH of the reaction medium was 5.5 at the beginning of the reaction and decreased during the process to 1 due to the acids formation [38-40]. The process was extremely slow and inefficient in the absence of catalyst and alcohol. Organic peroxides were formed in small amounts as the only products.

The oxidation process was more effective in the presence of vanadium catalysts and aliphatic alcohols. The conversion of 2-methylfuran increased by more than 30 times, the total yield of the oxidation products increased by more than 60 times and the duration of the process decreased by 15 times. At the same time, the reaction rate and yields of the products reached the highest values during oxidation in an acidic medium.

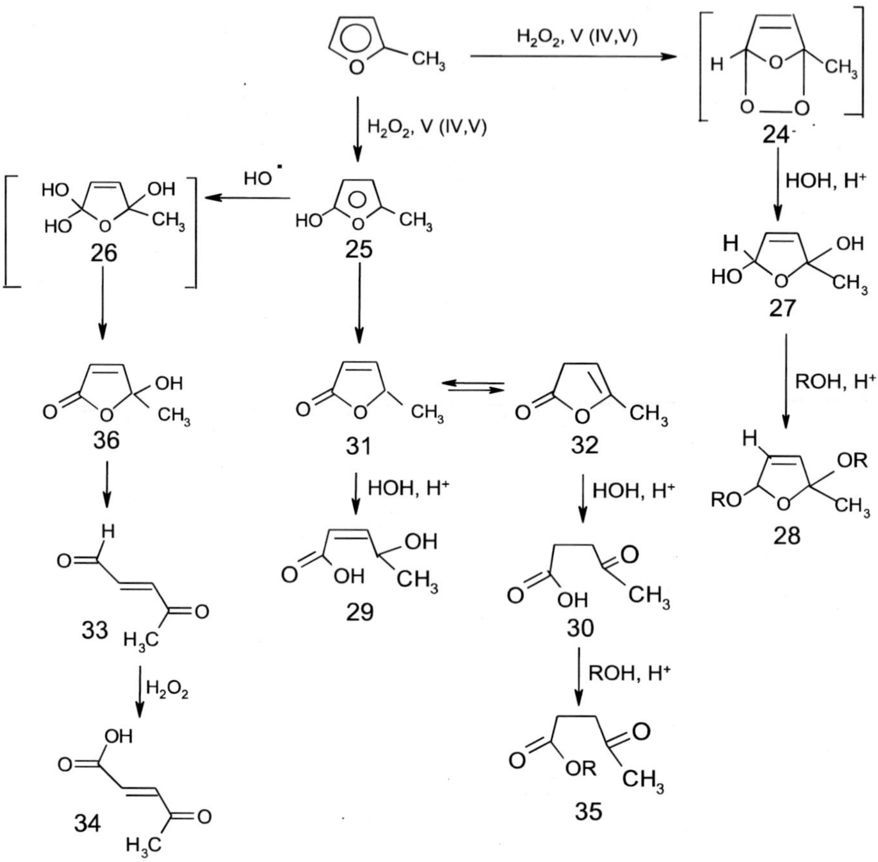

Scheme 1.4.

The increase in temperature from 20 to 25°C reduced the process duration to 1.5-2 h, increased the conversion of 2-methylfuran to 80% and increased the total yield of oxidation products to 70%. Further temperature rise resulted in unproductive consumption of H₂O₂ and undesirable deep transformations of the target products.

The type of vanadium catalyst also significantly affects the overall yield of oxidation products and the duration of the process. Thus, in the presence of $VOSO_4$, the yield of products was greater than in the presence of V_2O_5, although in the latter case the reaction duration was reduced.

It is significant that the reaction of 2-methylfuran with H_2O_2 in aqueous alcohols is 2-3 times faster than the oxidation of furan. This can be explained by the higher reactivity of alkyl substituted furans under these conditions.

The composition of 2-methylfuran oxidation products was studied by chromatography-mass spectrometry. The oxidation of 2-methylfuran in aqueous-alcoholic media was found to lead to organic peroxides as intermediates, which are further converted into carboxylic acids, carbonyl compounds, cyclic acetals and lactones.

According to chromatomass spectrometry data, 5-methyl-2(5H)-furanone (31), 5-methyl-2(3H)-furanone (32), levulinic acid (30) and its ether 35, 2-methyl-2,5-dialkoxy-2,5-dihydrofurans (28), 5-hydroxy-5-methyl-2(5H)-furanone (36), β-acetylacrylic acid (34) and 4-hydroxy-2-pentenoic acid (29) are the final products of 2-methylfuran oxidation in the system H_2O_2 – vanadium catalyst – aqueous ethanol (Scheme 1.4).

According to the authors [39, 40], the formation of the main products of this reaction passes through endoperoxide, 2-methyl-5-hydroxyfuran (24) and 2-methyl-2,5-dihydroxy-2,5-dihydrofuran (25) in two main directions (Scheme 1.4). These directions suppose attack of the furan nucleus by OH·radicals and peroxovanadium complexes, respectively. The formation of hydroxyfuran 25 and then the hydroxylation 26 can be explained by the interaction of 2-methylfuran with hydroxyl radicals. The interaction of furan with peroxovanadium catalyst presumably leads to 2-methyl-2,5-dihydroxy-2,5-dihydrofuran 27 via the intermediate formation of ozonide 24.

Dialkoxydihydrofurans 28 are presumably formed by alkylation of 2-methyl-2,5-dihydroxy-2,5-dihydrofuran (27). Acids 29 and 30 are presumably formed by hydrolysis of lactone 31, as well as by hydrolytic and tautomeric transformations of intermediate 32. Acid 34 is assumed to be the product of the oxidation of ketoaldehyde 33. Esterification of acid 30 leads to ether 35.

Cis- and *trans*-isomers of compound 28 are homologues of compounds formed during furan oxidation with hydrogen peroxide [38]. This indicates a certain similarity of the processes of furan and 2-methylfuran oxidation in the conditions under consideration.

It is noteworthy that the directions of 2-methylfuran oxidation passing through the lactones 31 and 32 formation and leading to the formation of hydroxyacid 29, levulinic acid (30) and its ester 35 are predominant. This can be explained by the stabilization of hydroxymethylfuran 25 by alkyl substituent. Intermediate 25 is further transformed into products 29, 30, 34 and 35.

Thus, the methyl substituent in the α-position of the furan cycle significantly accelerates its oxidation by hydrogen peroxide [39, 40]. The accelerating effect of the electron-donor methyl substituent indicates the predominance of electrophilic reactions involving the furan cycle in the studied process.

The presented results of 2-methylfuran oxidation by hydrogen peroxide form the basis for new methods of methyl substituted dialcoxydihydrofurans, methylfuranones and ketoacids syntheses. These methods can be considered as a promising alternative to existing methods of obtaining mentioned compounds [38-40].

HOMOGENEOUS CATALYTIC REACTIONS OF FORMYLFURANS WITH HYDROGEN PEROXIDE

Reactions of 2-formylfurans (furan aldehydes) with hydrogen peroxide differ significantly in speed and composition of the main products from similar reactions of furans considered in Section 1.

Thus, the reaction mixture of furfural (2-formylfuran) (36) and aqueous 27% hydrogen peroxide in a molar ratio of 1:2.6 at an initial temperature of 20^0C without thermostating begins to warm up due to the reactions started. The temperature becomes 50^0C after 8.5 hours, and then quickly reaches 108^0C. The mixture boils, and the reaction becomes violent. At this point, the oxidation process passes through the maximum formation of organic peroxides. Then it is completed by the total consumption of furfural and the formation of maleic, succinic and formic acids as the main products [41]. In this reaction mode, the degree of furan conversion in 72 hours is only 9%. In this case, oxidation products were found only in trace amounts and could not be identified.

This shows that the furan cycle is not the first reaction center for the interaction of H_2O_2 with formylfurans. The electron density in the formylfuran cycle is significantly reduced due to the coupling of the electron acceptor carbonyl group with the π–system of the cycle. On the contrary, it is increased on the oxygen of the carbonyl group. As a result, the aromaticity of the formylfuran cycle is much higher than in the furan molecule.

The structure and features of reactions of furan, formylfuran and benzene cycles, as well as carbonyl compounds of benzene series were considered and compared in the review [42]. Considering the presented data it is expected that the initial stages of the reaction of H_2O_2 with formylfurans participate in their carbonyl group. Oxygen of this group completes the conjugation system and is the center with the highest electron density.

2-Formylfuran (furfural) is most widely studied among the furan aldehydes in reactions with H_2O_2. The history of reactions of compound 36 with H_2O_2 from 1899 to 1965 was considered in the review [42]. According to the review, furfural was oxidized by 27-35% hydrogen peroxide at 50-100⁰C. The main reaction products, depending on the conditions, were carboxylic acids: succinic acid (yield 20-50%), maleic acid (10-40%), fumaric acid (2-25%), crude β-formylacrylic acid (up to 25%), and, in all cases, formic acid. 2-Furoic acid was obtained in aqueous pyridine. Oxidation of furfural in aqueous alcohol in the presence of OsO_4 led to the formation of β-formylacrylic and succinic acids.

The mechanism of these reactions has not been studied, and the authors put forward the schemes of chemical transformations taking place in these oxidation processes. It is noteworthy that, according to most of these schemes, the oxidation process is carried out by hydroxyl radicals or atomic oxygen formed from H_2O_2. Further, these particles are either attached by the furfural diene system or attack the furan formed during oxidative decarbonation of aldehyde 36 [42]. It is obvious that the proposed oxidation mechanisms do not agree with the properties of the cycle in furfural and furan. They do not take into account the peculiarities of hydrogen peroxide decomposition and oxidation mechanism under the considered conditions. Their authors do not take into account the results of the reactions of aromatic carbonyl compounds with H_2O_2 and do not provide for the formation of intermediate peroxides.

The papers [41-43] present the basics of modern concepts of homogeneous reactions of furan aldehydes with hydrogen peroxide. They first reported the formation of furan hydroxyperoxides in these reactions. Peroxides 37 and 38 (Schemes 2.1 and 2.2) and nitrosubstituted peroxide were synthesized by oxidation of furfural and 5-nitrofurfural respectively with concentrated hydrogen peroxide. They were singled out individually and identified [44, 45]. Isolation of 2-formyloxyfuran (39) from the reaction mixture and determination of their hydrolysis and oxidation products were important for understanding the mechanism of the furfural oxidation [46]. It was first established that 2(5H)-furanone (16) and 2(3H)-furanone (40) are among the main products of furfural oxidation by aqueous H_2O_2, and their homologues are formed during the oxidation of 5-methylfurfural [47]. Tautomeric, oxidative and hydrolytic transformations of lactones 16 and 40 explain the formation of some oxo- and dicarboxylic acids in these reactions (Scheme 2.1).

Reaction of furfural with hydrogen peroxide in water, ethanol and dioxane differ in the quantitative composition of the products, but peroxides are formed in each case. These peroxides are most stable in the oxidation of furfural in alcohol and dioxane (Table 2.1). The greatest formation of furanones 16 and 40 is achieved by oxidation of aldehyde 36 in water. In alcoholic solutions furan-2-carboxylic acid (41) is formed in a noticeable amount.

These facts allowed us to offer experimentally proven and theoretically (see below) reasonable Schemes 2.1 and 2.2, which describes the processes in the system furfural – H_2O_2 – water or water-organic solvent [48].

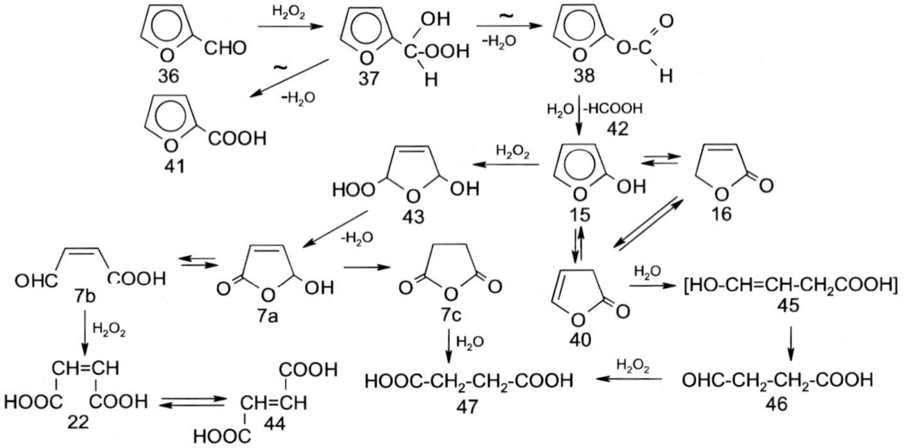

Scheme 2.1.

Table 2.1. Composition of products of the reaction between furfural and 30% H_2O_2 in different solvents[a]

Compound	Yield, mol% (with respect to furfural)		
	water	ethanol	dioxane
	180 min	210 min	440 min
Furfural α-hydroxyhydroperoxide (37)	10.8	70.1	98.0
Maleic acid (22)	6.6	0.8	Traces
β-Formylacrylic acid (7a and 7b)	10.8	3.1	–
Fumaric acid (44)	0.2	0.6	–
β-Formylpropionic acid (46)	4.0	1.0	–
Succinic acid (47)	20.0	3.0	–
2-Furoic acid (41)	1.3	8.1	–
Formic acid (42)	86	20	Traces
2(3H)-Furanone (40) and 2(5H)-furanone (16)	40	12	–

[a] [1] : [H_2O_2] = 1 : 2, 60 ± 1°C, conversion of furfural at the given reaction duration 70%.

It is noteworthy that the pH of the reaction medium in the oxidation of formylfurans in the presence of water is reduced to 1-2 as a result of the acids formation, primarily formic acid (42). The resulting acids act as catalysts in most of the stages presented in Scheme 2.1. Thus, the oxidation of furfural and other furan aldehydes by hydrogen peroxide has an acid-autocatalytic character [48, 49].

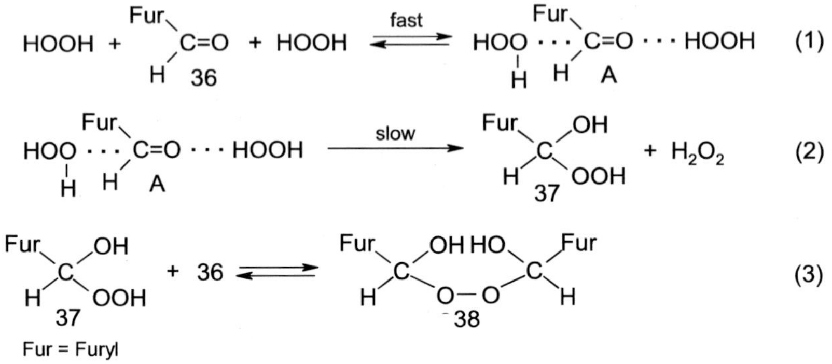

$$HOOH + \underset{36}{\underset{H}{Fur}\!\!>\!\!C=O} + HOOH \underset{}{\overset{fast}{\rightleftarrows}} \underset{A}{HOO\cdots \underset{H}{\underset{H}{Fur}\!\!>\!\!C=O}\cdots HOOH} \quad (1)$$

$$\underset{A}{HOO\cdots \underset{H}{\underset{H}{Fur}\!\!>\!\!C=O}\cdots HOOH} \xrightarrow{slow} \underset{37}{\underset{H}{Fur}\!\!>\!\!C\!\!<\!\!\underset{OOH}{OH}} + H_2O_2 \quad (2)$$

$$\underset{37}{\underset{H}{Fur}\!\!>\!\!C\!\!<\!\!\underset{OOH}{OH}} + 36 \rightleftarrows \underset{38}{\underset{H}{Fur}\!\!>\!\!C\!\!<\!\!\underset{O-O}{OH}\underset{}{HO}\!\!>\!\!C\!\!<\!\!\underset{H}{Fur}} \quad (3)$$

Fur = Furyl

Scheme 2.2.

The study of the furan aldehydes containing substituents which do not react with aqueous H_2O_2 [(5-methyl-furfural (48) and 5-nitrofurfural (49)] has allowed elucidation of the role of electron-donor (CH_3) and electron-acceptor (NO_2) groups [48, 50, 53]. These reactions occur through the formation of α-hydroxyperoxides of 5-methylfurfural (50) and 5-nitrofurfural (51).

The final products of oxidation of aldehyde 48 are methyl-substituted homologues of hydrofuranones 16 and 40 [(5-methyl-2(*3H*)-furanone (52) and 5-methyl-2(*5H*)-furanone (53)] and methyl-substituted structural analogs of acids 46 and 7b [acetylpropionic (levulinic) acid (30) and β-acetylacrylic acid (34)] (Scheme 2.1) [48, 50]. The reaction of aldehyde 48 with aqueous H_2O_2 does not yield 5-methyl-2-furanoic acid (54) in the specified conditions.

The oxidation of aldehyde 49 with 28% aqueous hydrogen peroxide diluted with dioxane gives exclusively 5-nitro-2-furanoic acid (55) in a quantitative yield.

As can be seen, Scheme 2.1 is consistent with the composition of intermediate and final oxidation products of furan aldehydes both with electron-donor substituents (for example, the 5-methylfurfural reaction) and with electron acceptor substituents (for example, the 5-nitrofurfural reaction).

The mechanism of homogeneous reactions of furan aldehydes with H_2O_2 in aqueous media and Scheme 2.1 are confirmed by the results presented in [48-51]. These discussed below results were obtained in the study of the formation and transformation of furan hydroxyperoxides (types 37 and 38) as well as acid-base catalysis and rearrangements and tautomeric transformations of products of formylfurans oxidation by hydrogen peroxide.

Furfural α-hydroxyhydroperoxide (37) is very unstable and explodes at room temperature in pure form [45]. In view of this, peroxide 37 has been prepared and studied in the form of a solution in n-butanol [51, 52]. Under these conditions oxidation of aldehyde 36 is fast enough, and the peroxide compounds are exclusively accumulated in the solution at certain moment. Besides peroxide 37, furfural α,α'-dihydroxyperoxide (38), a product of the reaction between aldehyde 36 and peroxide 37, is also formed (Scheme 2.2).

Kinetics of oxidation of furfural in n-butanol with 30% hydrogen peroxide prepared by dilution of 90% H_2O_2 with n-butanol [51, 52] has revealed the pseudo first rate order with respect to furfural and the second rate order with respect to H_2O_2, $i.\ e.$ overall the third reaction rate order. This has allowed description of the formation of peroxides 37 and 38 by Scheme 2.2.

In view of the kinetic data, the rate of the process in Scheme 2.2 is determined by stage 2. The formation of complex A (stage 1) is in agreement with the activation parameters of the reaction between furfural and H_2O_2 in n-butanol [51] and suggest the association between the carbonyl group of furfural and H_2O_2 molecules [53]. The third reaction rate order in the reaction of nucleophilic addition at the carbonyl group for synchronous addition of the nucleophile and electrophile (proton), which is the case of the considered process of attaching H_2O_2 to the carbonyl group of aldehyde 36 to form peroxide 37 as well. In the presence of acids, the reaction order is first in H_2O_2 and first in acid [51, 52].

It is notable that the formation of α-hydroxyhydroperoxides of furan aldehydes during oxidation of these aldehydes with hydrogen peroxide in dioxane (when exclusively peroxides are formed, Table 2.1) is accelerated by electron-donor (CH_3) as well as electron-acceptor (NO_2) substituents [51]. This is in line with the reaction through the formation of complex A (Scheme 2.2).

The oxidation of furfural in water gives exclusively peroxide 37, whereas peroxide 38 is formed in trace amount [54]. In this case, the formation of peroxide 37 is accompanied by its decomposition into furanones and acids (Scheme 2.1 and Table 2.1). At the same time, the third reaction rate order is preserved [51]: the first with respect to furfural, the first with respect to H_2O_2, and the first with respect to formic acid formed via hydrolysis of ester 39 and acting as a protogen in the formation of complex A (Scheme 2.2). This fact confirms that the formation of α-hydroxyhydroperoxide 37 limits the overall rate of the multistage reaction between furfural and H_2O_2 in an aqueous medium.

The products of α-hydroxyhydroperoxide 37 transformations in water have been compared with these obtained upon addition of water to its solutions in *n*-butanol, ethanol, and dioxane. The prepared aqueous-organic solutions containing 25% of water have been heated at 60°C until complete decomposition of the peroxides [51, 54]. A mixture of acids and hydrofuranones is formed in water (Table 2.1). In the case of the alcohols and dioxane with addition of water, qualitative composition of the products is the same in the specified conditions but 2-furoic acid (41) predominates among the acids (yield up to 60%), the yield of formic acid being reduced.

It is notable that the oxidation of 5-methylfurfural (48) with aqueous hydrogen peroxide does not give 5-methyl-2-furoic acid (54). At the same time, the oxidation of aldehyde 48 in a more basic solvent (water : triethylamine = 1 : 1) yields acid 54 as the major product [55].

The study of transformations of α-hydroxyhydroperoxide 37 in pyridine and triethylamine at 24–25°C has revealed that it is completely consumed during 15–20 min with quantitative formation of acid 41. The decomposition of peroxide 37 follows the first-order kinetics, and the rate as well as activation parameters in both solvents are almost equal

(Table 2.2). The activation parameters point at the formation of a highly polar transition state during the transformation of peroxide 37 into acid 41. Hence, the transformation of furfural and 5-methylfurfural α-hydroxyhydroperoxides into the corresponding 2-furoic acids is the major pathway in the basic solvents.

Table 2.2. Parameters of transformation of furfural α-hydroxyhydroperoxide (c_{start} = 0.1 mol/L) into 2-furoic acid in basic medium

Solvent	T, 0C	k.10-2, s-1	Ea, kcal/mol	A.10-5, s-1	ΔSa, e. u.
Pyridine	24	1,65	10.0	3.7	-26
Triethylamine	25	1.75	10.0	3.9	-25

In contrast to peroxide 37, 5-nitrofurfural α-hydroxyhydroperoxide (51) is relatively stable. It has been prepared via oxidation of 5-nitrofurfural (49) with 35% H_2O_2 in a water–dioxane medium at 45°C in 95% yield [55]. Peroxide 51 is a stable crystalline compound of mp 79–80°C. Upon 2 h heating in water at 70°C it is quantitatively transformed into 5-nitro-2-furoic acid (55). It is notable that the oxidation of aldehyde 49 with 30% H_2O_2 gives peroxide 51 as the only intermediate, which is further transformed into acid 55 in 98% yield [51].

Comparison of the reactions of H_2O_2 with aldehydes 36 and 48, and furan ketones such as 2-acetylfuran (56), and 5-methyl-2-acetylfuran (57) has shown the identity of their products and oxidation mechanisms [51]. Chromatographic analysis of the reaction mixture has revealed the presence of α-hydroxyhydroperoxides with structure and properties analogous to these of peroxides 37 and 50 formed from furfural and 5-methylfurfural, respectively. The products of deep oxidation of ketones have been isolated as well: 2(3*H*)-furanone (40), 2(5*H*)-furanone (16), 5-methyl-2(3*H*)-furanone (52), and 5-methyl-2(5*H*)-furanone (53); β-formylacrylic (7), maleic (22), succinic (47), β-acetylacrylic (34), and β-acetylpropionic (30) acids. The furanones 52 and 53 are major products of the ketones oxidation. The oxidation of furan ketones is accompanied by quantitative formation of acetic acid. Esters of 2-furoic acid or 5-methyl-2-furoic acid are not formed.

The oxidation of furan ketones is slower in comparison with furan aldehydes, as expected in view of the effect of the methyl group increasing the electron density at the carbon atom of the carbonyl group and sterically hindering the nucleophilic attack of this atom by hydrogen peroxide.

The discussed data have suggested the concept of the rearrangements of furan α-hydroxyhydroperoxides in homogeneous reaction mixtures containing the furan aldehyde, H_2O_2, and solvent. The formation of such hydroxyhydroperoxides occurs through the formation of type A complexes (Scheme 2.2), with hydrogen peroxide acting simultaneously as nucleophile and electrophile. In the case of reactions of furfural and 5-methylfurfural in water, the forming organic acids (mainly formic one) begin to act as protogen in complexes A (Scheme 2.2). α-Hydroxyhydroperoxides are protonated (by hydrogen peroxide and then with the formed acids) at the α-oxygen atom of the peroxide fragment (Scheme 2.3). This enhances the polarization of the peroxide bond and additionally reduces the electron density at its α-oxygen atom. The formation of an intramolecular hydrogen bond between the hydroxyl and peroxide groups with the formation of a sufficiently stable five-membered cycle is also probable (complex B, Scheme 2.3).

The Baeyer–Villiger rearrangements of α-hydroxyhydroperoxides of furfural, furan aldehydes with electron-donor substituents in the furan ring (as exemplified by 5-methylfurfural), and furan ketones with an alkyl substituent or without other substituents in the ring proceeds with the migration of furyl radicals to the electron-deficient α-oxygen atom of the peroxide bond (Scheme 2.3). This is a distinct feature of these rearrangements in comparison with the rearrangements of similar peroxides of benzaldehyde and benzene ketones. This fact can be explained by the enhanced electron density in the furan ring of furan peroxides owing to the conjugation with the heteroatom as well as lability of the π-electron system not typical of the benzene cycle. Hence, rearrangement of the listed furan aldehydes involve the overlapping of the *p*-orbital of the electron-deficient α-oxygen atom of the peroxide bond in the molecule of peroxides 37 and 50 with the σ-orbital of the adjacent C–C bond (furyl), bonding electrons of which act as donor (transition state C, Scheme 2.3).

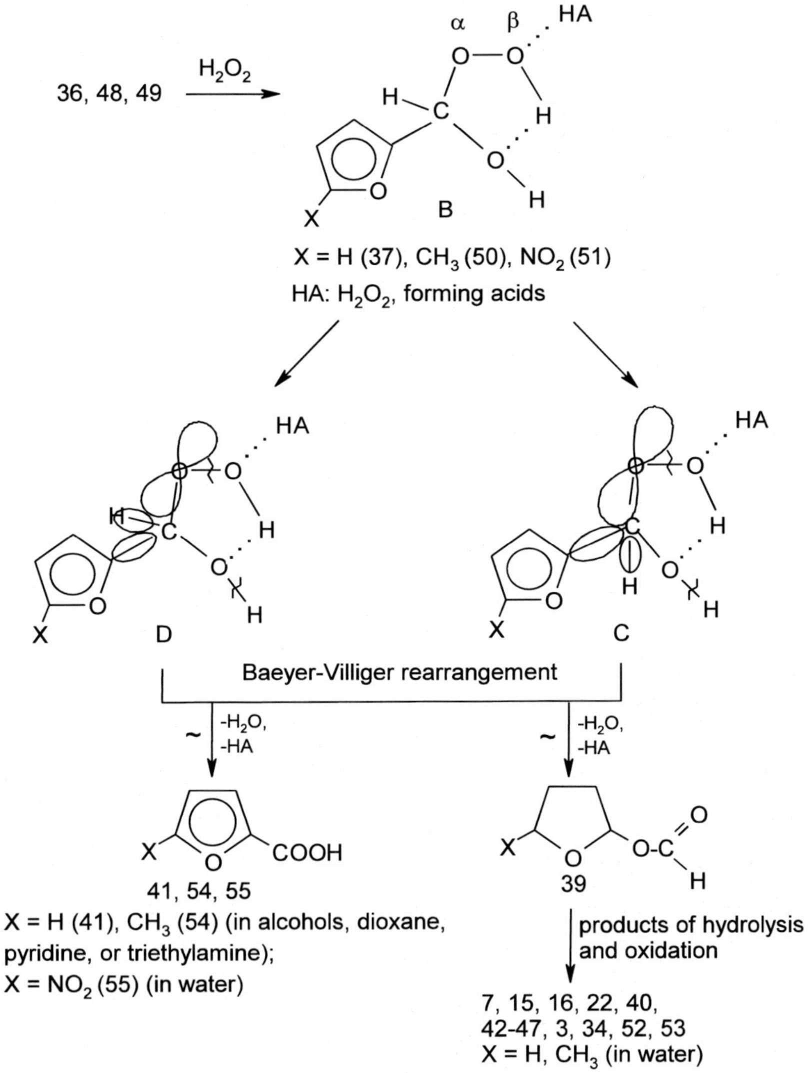

36, 48, 49
H_2O_2
α
β
HA
H
B
X
X = H (37), CH_3 (50), NO_2 (51)
HA: H_2O_2, forming acids
D
C
Baeyer-Villiger rearrangement
$-H_2O$,
-HA
COOH
41, 54, 55
X = H (41), CH_3 (54) (in alcohols, dioxane, pyridine, or triethylamine);
X = NO_2 (55) (in water)
39
products of hydrolysis and oxidation
7, 15, 16, 22, 40,
42-47, 3, 34, 52, 53
X = H, CH_3 (in water)

Scheme 2.3.

Table 2.3. The effect of the medium pH on the formation of 2(5H)-furanone (12) and carboxylic acids 4 and 15 in the furfural– H_2O_2–H_2O system[a]

pH	Furfural conversion, %	Yield of oxidation products, %		
		16	41	47
0	100	38	–	40
1	100	50	–	40
2	98	40	–	40
3	80	22	–	43
4	65	13	–	45
5	52	–	30	50
6	48	–	50	40
7	50	–	65	20
7.5	58	–	45	4

[a] [Furfural] : [H_2O_2] = 1 : 10, 60°C, 4 h.

(A-X) basic solvent (alcohols, dioxane, pyridine, or triethylamine)

Scheme 2.4.

If the furan ring contains an electron-acceptor group (as in the case of 5-nitrofurfural), the electron density in the cycle is sufficiently reduced, and the formation of such transition state is impossible. In this case, hydrogen atom migrates to the electron-deficient atom of peroxide 51 through transition state D (Scheme 2.3), in which the bonding electrons of the adjacent C–H σ-bond act as donors, and 5-nitro-2- furoic acid (55) is the product of the Baeyer–Villiger rearrangement. This pathway of rearrangement of peroxide 37 during oxidation of furfural (36) in an aqueous medium is insufficient, being absent for the reaction of 5-methyl-furfural. However, 2-furoic acid (41) and 5-methyl- 2-furoic acid (54) are the major products of oxidation in the reactions of these aldehydes with

H_2O_2 in more basic organic and aqueous-organic media (alcohols, pyridine, or triethylamine) [55].

pH of the reaction medium also affects the direction of the rearrangement of furfural α-hydroxyhydro-peroxide (37) [49]. As shown in Table 2.3, the reaction of aldehyde 36 with H_2O_2 in an aqueous medium, the formation of acid 41 starts only at pH 5, being the most pronounced at pH 7. In acid aqueous medium (pH 0–4) furanone 16 and acid 47 are mainly formed. The presence of protons first favors the formation of peroxide 37 and then induces the polarization of the peroxide bond reducing the electron density at its α-oxygen atom. This, on the one hand, facilitates the rearrangement of peroxide 37 and, on the other hand, favors the formation of formyloxyfuran 39. In an aqueous medium, the latter is hydrolyzed into hydroxyfuran 15, which undergoes further transformations (Scheme 2.1).

The predominant formation of 2-furoic acid (41) via rearrangement of furfural α-hydroxyhydroperoxide (37) observed in basic organic media can be explained by the formation of hydrogen bonds between the molecules of the organic solvent and protons of the furan ring of the peroxide. The formation of such bonds upon dissolution of furfural in dioxane and DMSO has been confirmed by means of ^1H NMR spectroscopy [56]. Similarly, the formation of complexes E and F involving peroxide 37 (Scheme 2.4) during the reaction of furfural with H_2O_2 in basic media can be assumed. The formation of dioxane – 5-nitrofurfural α-hydroxyhydroperoxide (51) complexes has been observed during the oxidation of 5-nitrofurfural with hydrogen peroxide in dioxane [56].

The formation of such bulky structures sterically hinders the migration of furan ring to the electron-deficient α-oxygen atom of the peroxide bond through complex C (Scheme 2.3) and, hence, the formation of formyloxyfuran 39. In this case, migration of proton through complex D and formation of acid 41 are favorable.

The discussed data show that the Baeyer–Villiger rearrangement of a series of furan α-hydroxyhydro-peroxides (which have been isolated in the reactions of furan aldehydes with H_2O_2) exhibits specific features in comparison with the rearrangements of similar benzene-type peroxides.

This occurs owing to the differences in the structure and properties of the furan and benzene rings [41].

Further transformations of formyloxyfuran 39 and formyloxybenzene formed via rearrangement of the corresponding α-hydroxyhydroperoxides are significantly different as well. The products of hydrolysis of formyloxybenzene (formed via the oxidation of benzaldehyde with hydrogen peroxide) are formic acid (42) and phenol [42]. Hydrolysis of formyloxyfuran 39 yields acid 42, 2-hydroxyfuran (15) and its tautomeric forms (hydrofuranones 16 and 40) (Scheme 2.1). Electronic structure and properties of the furan ring of substituted furan 5 strongly differing from these of the benzene ring. This is the reason for other differences in the transformations of formyloxybenzene and formyloxyfuran in the reaction systems where these are formed via the Baeyer–Villiger rearrangement.

Products of hydrolysis of ester 39 are acid 42, lactones 16 and 40, and semialdehyde of succinic acid (46) (Scheme 2.1). Compound 46 is formed by hydrolysis of lactone 40 [47]. The hydrolysis is relatively fast ($k = 13.0 \times 10^{-3}$ min^{-1}). Lactone 16 is stable and has been isolated in a yield up to 40% during oxidation of furfural with hydrogen peroxide in an aqueous medium [50]. The freshly prepared furanone 16 contains up to 20% of furanone 40 as the admixture; the latter is transformed into lactone 16 upon heating or 2 months storage at room temperature [47]. High-resolution NMR spectrum recorded for the purified lactone 16 contains relatively weak signals at δ 4.17, 5.80, 6.58, and 6.95 ppm [47]. They are attributed to the presence of 2-hydroxyfuran (15), the tautomeric form of furanone 16. This has been further confirmed by the observation of a weak band at 3500 cm^{-1} in the IR spectrum of lactone 16, typical of the OH group of aromatic enols. The enol 15 has been observed also in the products of the oxidation by means of chromato–mass spectrometry.

The available data have allowed the explanation of the mechanism of the lactones and acids formation in the furfural–H_2O_2–H_2O reaction system. The readily proceeding hydrolysis of ester 39, in contrast to this of formyloxybenzene, is due to the formation of unstable enol 15 from ester 39.

Positive effect of the hydroxyl group conjugation with the π-system of the furan ring occurs in the molecule of hydroxyfuran 15. As a result, the electron density in the cycle is enhanced, its aromaticity is reduced, and the diene system reactions become characteristic of it. This leads to the easy mobility of the hydroxyl group hydrogen to the position 3 of the furan cycle (adjacent to the hydroxyl group) and to the opposite position 5, explaining the tautomeric trans-formation of enol 15 into lactones 16 and 40 (Scheme 2.1). Partial preservation of the conjugated bonds system in the molecules of lactones contributes to these transformations. Such transformations have been described in the literature [57]. At the same time, as mentioned above [47], lactone 16 contains minor amount of the enol 15. This is a typical case when the enol double bond is conjugated with other double bonds [58], as in the molecule of intermediate 15.

Hydrofuranone 40 is readily hydrolyzed into acid 45 which exists in the stable tautomeric form of aldehyde-acid 46 (Scheme 2.1). The formation of acid 46 from 2(3H)-furanone (40) has been confirmed experimentally [47]. The aldehyde-acid 46 is oxidized with hydrogen peroxide in the reaction mixture into succinic acid 47, which has been isolated from the products of the furfural reaction with H_2O_2 in up to 40% yield [50].

In contrast to lactone 40, compound 16 is stable in acidic aqueous media, including the conditions of furfural oxidation with hydrogen peroxide in water. It does not undergo acid hydrolysis and is not oxidized by hydrogen peroxide in acid media [50].

The product of oxidation of tautomer 15 is β-formylacrylic acid (7) existing in the reaction medium in two tautomer forms 7a and 7b. It is formed via the addition of H_2O_2 at the diene system of 2-hydroxyfuran (15) through the formation of hydroperoxide 43 (Scheme 2.1). Its yield is low upon oxidation of furfural 36 with hydrogen peroxide in water in the absence of a catalyst (Table 2.1). However, the oxidative ability of the medium is enhanced when the reaction of aldehyde 36 with H_2O_2 is performed in the presence of selenium, vanadium, molybdenum, or niobium compounds, and the oxidation of hydroxyfuran 15 becomes competitive to its tautomeric transformations into lactones 16 and 40. Under these conditions, acid 8 is formed in significant yield [59, 60].

Acid 7b is further oxidized into maleic acid (22), which it slightly isomerized into fumaric acid (44). The ease of transformation of hydroxyfuran 15 into stable products 7a, 7b, 16 and further 22 and 47 favors the shift of the equilibrium of hydrolysis of ester 39 towards the intermediate 15 and the products of its tautomeric and oxidative transformations. The formed acids 7b, 22, 42, and 47 act as catalysts at all the stages of the reaction in an aqueous medium shown in Scheme 2.1.

Oxidation of benzene with hydrogen peroxide stops at the formation of phenol, the phenol–ketone tauto-meric equilibrium of which is shifted towards phenol due to aromaticity of the benzene ring [58]. This fact rules out any further transformations of phenol analogous to these shown in Scheme 2.1 for furan aldehydes. Hence, it is the keto-enol tautomerism of 2-hydroxyfuran (15), intermediate of furfural oxidation with hydrogen peroxide, which determines the dramatic difference of this process from benzaldehyde oxidation with H_2O_2.

In the furfural–H_2O_2–H_2O reaction system the tautomeric transformations involving the proton transfer are observed also for the products of hydroxyfuran 15 oxidation (Scheme 2.1). β-Formylacrylic acid (7) has been isolated from the reaction medium in two forms: open carbonyl (7b) and cyclic hemiacetal (7a) ones, their ratio being changed in the course of the reaction. Initially, when the reaction medium pH is close to neutral, the carbonyl form 7b is predominant. As the medium becomes more acidic, the ratio is changed in favor of the cyclic form 7a which is prevailing in the final products. pH of the reaction medium is reduced from 6.5 to 1 due to the formation of acids 6, 7b, 42, and 47 (Scheme 2.1). To explain the observations, the equilibrium between compounds 7a and 7b has been studied by means of polarography [61]. The reduction has been performed using a dropping mercury electrode with m 2.01 mg/s and t_1 3.41 s. Only the 7b form can be reduced under the experiment conditions, which allows monitoring the ratio of the forms by measuring the current of reduction.

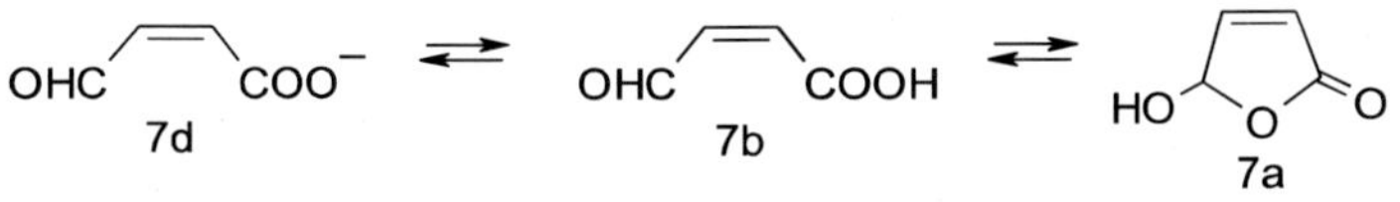

Scheme 2.5.

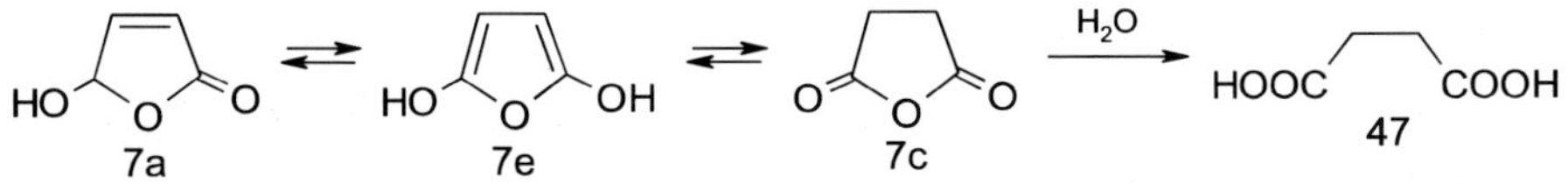

Scheme 2.6.

It has been found that β-formylacrylic acid exists in almost exclusively open neutral 7b and anionic 7d forms in basic to weakly acidic media (Scheme 2.5). At pH 0–4, 30–60% of it is in the cyclic form 7a. The *trans*-form of acid 7 has not been observed.

The obtained data on the pH effect on the cycle-chain tautomerism of acid 7 are in agreement with the behavior of other acylacrylic acids.

The observed partial transformation of the hemiacetal form 7a into succinic acid (47) [62] has not been known before. Heating of a neutral aqueous solution of compound 7a (c = 1.2 mol/L) at 60°C has afforded acid 47 with 3% yield after 3 h and 13% yield after 24 h. At pH 9–10 and 70–80°C, hemiacetal 7a is completely transformed into acid 47 within minutes; the latter has been isolated as such and in the form of diethyl ester. Scheme 2.6 of the transformations has been suggested.

Hemiacetal 7a containing an oxo group is in equilibrium with the enol form 7e which can have a keto form 7c, succinic anhydride. Hydrolysis of the latter gives acid 47. In the basic media the equilibrium is shifted from the form 7a towards form 7c due to the hydrolysis of the latter into acid 47. Hence, it can be suggested that acid 47 is partially formed from hemiacetal 7a in the furfural–H_2O_2–H_2O system (Scheme 2.1) when the medium acidity has not yet significantly increased.

The above discussion has revealed that most of the stages of the reactions between furan aldehydes with hydrogen peroxides occur through rearrangements and tautomeric transformations of furan and hydrofuran compounds 7a, 15, 16, 37, and 40. The predominant formation of 2-furoic acids or hydrofuranones 7a, 16, and 40 and acids 7b, 22, and 47 in this complex oxidation process is directed by the substituent in the furan cycle, nature and basicity of the solvent, and acid-base properties of the medium.

Methodological approaches to the synthesis of these compounds are given in section 3.

Besides the considered oxidation processes, homogeneous reactions of furfural with H_2O_2 in the presence of compounds of V and VI group elements soluble in the reaction system have been described [52, 59, 60, 63].

The addition of catalytic amount of a soluble compound of V^{+4}, V^{+5}, Nb^{+2}, Nb^{+5}, Mo^{+6}, Cr^{+6}, or Se^{+4} to the furfural–H_2O_2–water reaction system immediately affords the peroxo complexes of these elements, as confirmed in the study of the model systems by means of spectroscopy and chromatography. These complexes vigorously react with the carbonyl group of furfural with the formation of the corresponding furyl-containing organometal ozonides which are likely transformed into 2-formyloxyfuran (39) (Scheme 2.7).

In addition to ozonides, several peroxides of unknown structure are formed at the reaction intermediate stages. Perhaps these peroxides are the intermediates of ether 39, hydroxyfuran 15 and oxoacids 7b and 46 oxidation. Furfural α-hydroxyhydroperoxide (37) is not formed in these reactions.

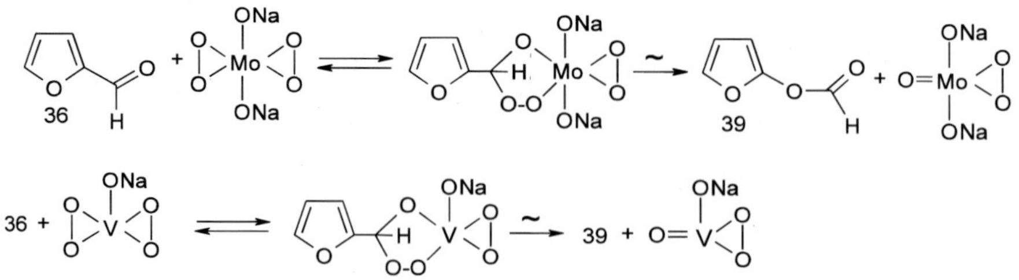

Scheme 2.7.

The reactions are the fastest in the presence of Se^{+4} or V^{+5} compounds, being the slowest in the presence of niobium compounds [60]. The rates of the reaction in the presence of Nb^{+2} and Nb^{+5} were equal and close to that in the absence of any catalyst. This observation can be explained by the negative oxidation potential of niobium in its compounds, regardless of the oxidation state. The final composition of the products of the catalytic oxidation of furfural in the presence of V^{+4}, V^{+5}, Mo^{+6}, Cr^{+6}, or Se^{+4} differs from that under the acid autocatalysis conditions [60] (Table 2.4). In the presence of V^{+4} or V^{+5}, β-formylacrylic acid (as a mixture of two tautomeric forms 7a and 7b) is predominantly formed, its yield being insignificant in the absence of the catalyst. The reactions in the presence of Cr^{+6} give mainly acids 7 and 47, whereas the presence of Mo^{+6} triggers the formation of furanone 16 and hydroxyacids (tartaric 58 and malic 59 ones). Niobium compounds do not significantly affect the composition of the major products of the furfural oxidation as compared to the catalyst-free process (Table 2.4).

Table 2.4. Composition of major products formed in a homogeneous reaction mixture furfural–H_2O_2–water–catalyst upon complete decomposition of peroxide compounds[a]

Catalyst	Yield, %[b]					
	7a+7b	16	22	47	58	59
Acid autocatalysis	4	22	8	48	–	1
Nb_2O_5	1	29	12	17	–	–
$Nb(CH_3COO)_2$	1	34	8	14	–	–
$NaVO_3$[c]	33	8	9	15	3	4
$Na_2Cr_2O_7$	22	25	9	10	4	1
Na_2MoO_4	7	33	1	11	8	27

[a] [1] : [H_2O_2] : [catalyst] = 1 : 3.2 : 0.05, 60 ± 1°C.
[b] A mixture of β-formylpropionic, malonic, and oxalic acids is also formed (overall yield 20% in the presence of Nb^{+2} and Nb^{+5}, 10% in the presence of V^{+5}); 1–6% of malonic acid is formed in the presence of Mo^{+6} and Cr^{+6}.
[c] The catalyst loading was reduced tenfold due to the vigorous reaction.

The oxidation of furfural with 5% H_2O_2 at 50 ± 1°C in the presence of SeO_2 (furfural : H_2O_2 : catalyst molar ratio 1 : 3 : 0.05) leads to the predominant formation of β-formylacrylic acid (7) (as a mixture of tautomers 7a and 7b, yield 60%) along with maleic (22) and fumaric (44) acids (total yield 22%), succinic (47) and β-formylpropionic (46) acids (total yield 8%), and 2(5H)-furanone (16) (yield 8%) [52, 59, 60].

In all the mentioned cases, formic acid 42 is formed with a quantitative yield.

The well-known epoxidizing ability of molybdenum peroxo complexes and characteristic transformation of vanadium peroxo complexes into the complexes with singlet oxygen $V^{+5}(O_2)$ determines the differences in the pathway and mechanism of the stage the ester 39 and hydroxyfuran 15 oxidation as compared to the acid autocatalysis conditions (Scheme 2.8), which leads to the formation of compounds 7, 58, and 59 in significant amounts in this reactions (Table 2.4) [60].

It should be noted that considered homogeneous oxidation of formylfuran by H_2O_2 in water and homogeneous aqueous-organic media in the conditions of acid autocatalysis has no features of radical reactions. This is proved by the following facts. Inhibitors of chain radical reactions do not affect their speed and composition of products, and styrene added to oxidates is not polymerized [51].

In addition, it was previously established that hydroxylation of furan cycles by OH˙radicals, which could be formed from H_2O_2 and participate in these reactions, leads to different kinetic results compared to the oxidation of furfural with hydrogen peroxide in the studied conditions [64].

Comparison of the reactions of 2-formylfurans with hydrogen peroxide in the presence of vanadium compounds with similar reactions of furans (Section 1) shows that under these conditions the oxidation of furan and methylfuran is much slower than the oxidation of furfural and 5-methylfurfural [65].

In the initial stages of formylfuran oxidation, the reactionary center is the carbonyl group, but not a furan core. This explains the difference in the chemical behavior of furans and formylfurans. At later stages (Schemes 2.1 and 2.8) 2-formyloxyfuran (39) is formed. Intermediate 39 is easily hydrolyzed to 2-hydroxyfuran (15) in the presence of water in the reaction mixture. The electron density in the furan cycle of compound 15 is significantly increased due to its conjugation with oxygen of the hydroxyl group. As a result, the aromaticity of the furan cycle of compound 15 decreases in comparison with furan, while its ability to split and react as diene increases. This determines the ease of subsequent reactions involving the furan cycle of compound 15 with the formation of products 7a, 7b and tautomeric forms of 2-hydroxyfuran – lactones 16 and 40 (Schemes 2.1 and 2.8).

A considerable acceleration of the reaction of the formylfurans with H_2O_2 after decomposition of peroxides 37 and formyloxyfuran and 2-hydroxyfuran formation is determined by these circumstances [51]. These facts also explain the greater activity of formylfurans in the reaction with aqueous hydrogen peroxide compared with furan and 2-methylfuran.

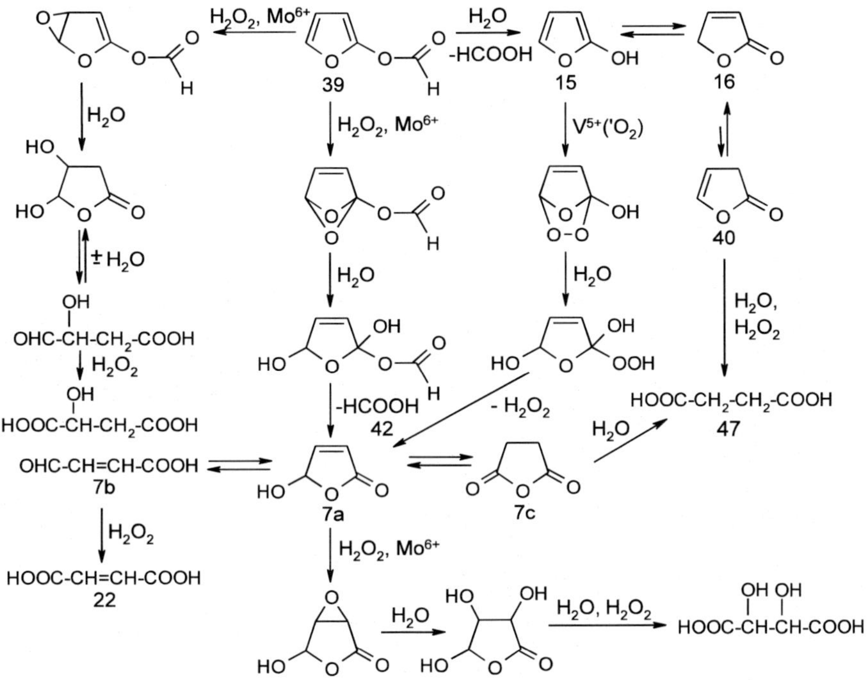

Scheme 2.8.

METHODS FOR THE SYNTHESIS OF FURAN AND HYDROFURAN COMPOUNDS BASED ON HOMOGENEOUS REACTIONS OF FURANS AND 2-FORMYLFURANS WITH HYDROGEN PEROXIDE

The influence of the furan compound structure and reaction factors on the composition of the products of furan, 2-methylfuran and furan aldehydes oxidation by hydrogen peroxide is considered in sections 1 and 2. Using these data, new methods of synthesis of a number of furan and hydrofuran compounds have been developed [51, 55]. The electron acceptor substituents in the cycle of furan aldehydes were found to contribute to the rearrangement of hydroxyhydroperoxides type 37 (Scheme 2.3, Section 2) in the corresponding 2-furoic acids [48]. With this in mind, methods for the synthesis of substituted 2-furoic acids 55, 60-62 have been developed (Scheme 3.1).

In all cases, the starting furan aldehyde was dissolved in dioxane with heating and slowly treated with 30% aqueous hydrogen peroxide; the reaction was carried out under vigorous stirring at 70°C until complete consumption of aldehyde. For the synthesis, a reagent molar ratio was 1:2. Acids were obtained in the yields of 92-98%.

As noted in section 2, the rearrangement of hydroxyperoxides of furfural and 5-methylfurfural in solvents more basic than water or at a pH of more than 5 mainly leads to the corresponding 2-furoic acids [48] (Scheme 2.3). With this in mind, two methods for the synthesis of 2-furoic (41) and 5-methyl-2-furoic (54) acids have been developed [55, 66].

The first procedure involves stirring a mixture of furfural, triethylamine, and 30% aqueous hydrogen peroxide in a molar ratio of 1:1:2 at 20-25°C until complete consumption of furfural. In this case, the yield of acid 4 was 92%. Following the second procedure, a solution of furfural in n–butanol was treated portionwise with a solution of sodium acetate in 30% aqueous H_2O_2 (1: H_2O_2 : AcONa = 1 : 2 : 0.5) over 2h. The reaction was carried out under stirring with simultaneous azeotropic distillation of water; the water distilled off was periodically returned into the reaction. After complete conversion of furfural, the solvent was removed, the residue was treated with hydrochloric acid, and the crystals of acid 4 was collected by filtration; yield of 41 was 72%. 5-Methylfuran-2-carboxylic acid is prepared in a similar way from 5-methylfurfural (48). The possibility to direct the reaction of furfural with H_2O_2 in the presence of alcohols and amines to produce acid 41 was used for developing the procedures to access esters and amides of this acid from aldehyde 36 without intermediate isolation of acid 41 [55, 67, 68] (Scheme 3.2).

Esters 63–65 are synthesized by oxidation of fufural with 5% solution of hydrogen peroxide in ethanol (prepared from 90% aqueous H_2O_2) in the presence of catalytic amounts of SeO_2. The reaction is carried out under vigorous stirring at 50°C using a molar ratio 1:H_2O_2:SeO_2 = 1:3:0.05 until complete consumption of furfural. Then the same amount of the catalyst is additionally added and the reaction mixture is refluxed for 5 h. This method gives 60–62% yields of the target esters.

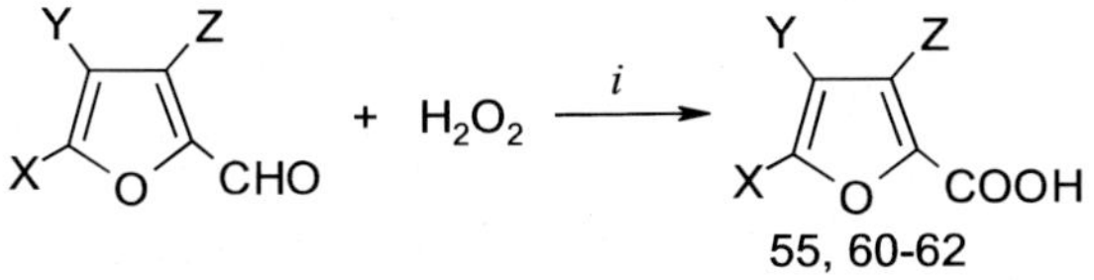

	X	Y	Z		X	Y	Z
55	NO_2	H	H	61	NO_2	H	Br
60	NO_2	Cl	H	62	COOH	H	H

Conditions: *i*. 70 °C, water/dioxane

Scheme 3.1.

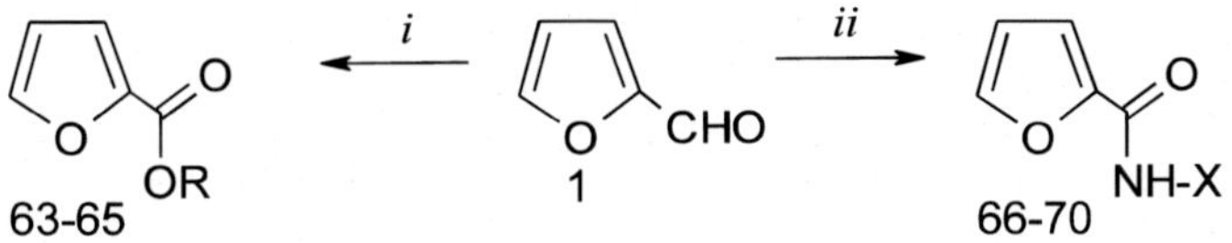

R = Et (63), Pr (64), Bu (65); X = Ph (66), 4-BrC_6H_4 (67), 4-HOC_6H_4 (68), $PhCH_2$ (69), H (70)
Reagents and conditions: *i*. H_2O_2, SeO_2, ROH.
ii. H_2O_2, aqueous EtOH, H_2N-X.

Scheme 3.2.

2–Furancarboxamides 66-69 are synthesized by the reaction of furfural with 30% solution of hydrogen peroxide in aqueous alcoholic solution catalyzed by primary aromatic amines (Scheme 3.2). The reactions are performed under stirring at 0–2°C using a molar ratio aldehyde 1 : H_2O_2 : amine = 1 : 2 : 1 until complete consumption of furfural.

The first representative of a homologous series of amides of acid 41, i.e., unsubstituted amide 70, was synthesized by oxidation of furfural with 30% solution of hydrogen peroxide in 25% aqueous ammonia in the presence of vanadium(IV,V) compounds [55, 69]. A stirred mixture of the reagents (a molar ratio of 1:H_2O_2:ammonium:vanadium compound = 1:3:5:0.005) is heated at 50–60°C until complete consumption of furfural. The yield of amide 70 reaches 84% [69]. As compared with the catalyst–free conventional synthesis of amide 70 in the system aldehyde 36–H_2O_2–NH_4OH, the presence of vanadium catalysts results in a two–fold increase in the target product yield and a decrease in the reaction time to 6 h.

The oxidation of furfural by aqueous H_2O_2 at pH above 8 in the presence of catalytic amounts of vanadium compounds was found to lead to the preferred formation of a new polyhydroxycarboxyfuranone 71 (Scheme 3.3) [49]. Model experiments have shown that the furanone 71 is the product of the transformation of 2-furoic acid formed during the oxidation of furfural under the above conditions at pH 7-8 [49].

For the synthesis of 3,4,5-trihydroxy-5-carboxy-2(3H,5H)-furanone (71), furfural is oxidized with 30% aqueous hydrogen peroxide in the presence of $VOSO_4$ at a molar ratio 1 : H_2O_2 : $VOSO_4$ = 1 : 5 : 0.005 at pH 8—9 and 60°C. The yield of product 71 is 68% [70, 71].

The reaction of furfural with either 15% or 30% hydrogen peroxide is performed at a molar ratio of 1: (2.0—2.2). The oxidation is carried out at 60–70°C until complete consumption of furfural. Then the oxidate is concentrated, the crystalline product is filtered off, and the filtrate is extracted with chloroform to give lactone 16 in 40% yield.

The crystalline product is succinic acid (47) formed from unstable lactone 40 upon its hydrolysis to β–formylpropionic acid (Scheme 2.1).

The latter compound is oxidized to acid 47 in 40% yield. Thus, under the described conditions the overall yield of compounds 16 and 47 is 80%.

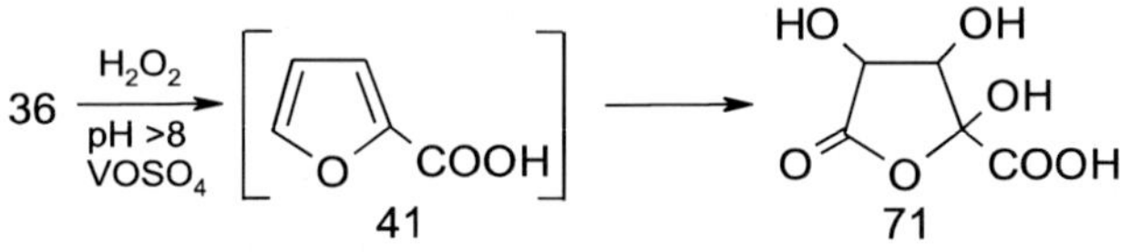

$$36 \xrightarrow[\substack{\text{pH} > 8 \\ \text{VOSO}_4}]{\text{H}_2\text{O}_2} \left[\begin{array}{c} \text{furan-COOH} \\ 41 \end{array} \right] \longrightarrow \begin{array}{c} \text{compound 71} \end{array}$$

Scheme 3.3.

Earlier, furanone 16 was synthesized by a multistep protocol from substituted butyrolactones, which in turn were prepared from poorly available substituted butanoic acids or vinylacetic acid as well as by hydrolysis of 2–acetoxy and 2–methoxyfuran. After our publication [72] appeared, Liu and co–workers [73] described synthesis of lactone 16 in 67% yield by oxidation of furfural with 30% hydrogen peroxide in heterogeneous system water–dichloroethane in the presence of sodium sulfate at 70°C with all other conditions being identical to the earlier published [72]. The total reaction time is 14 h.

2(5H)-Furanone (16) was also obtained by homogeneous oxidation of furfural with hydrogen peroxide in the presence of Nb(Ac)$_2$ [74]. A mixture of furfural, 38% aqueous H$_2$O$_2$, water and Nb(Ac)$_2$·4H$_2$O (a molar ratio 1 : 1 : 7 : 0.0015) was stirred at 60°C until all peroxides were completely converted. Product 16 was extracted with sulfur ether, washed with NaHSO$_3$ solution, dried with Na$_2$SO$_4$ and evaporated. The residue was distilled in a vacuum. The yield of furanone 16 was 64%.

According to recent reports [75-77], 2(5H)-furanone (16) is also obtained by oxidation of furfural with hydrogen peroxide in heterogeneous systems.

The oxidation with 30% hydrogen peroxide in a water–1,2-dichloroethane in the presence of Na$_2$SO$_4$ at 70°C, followed by boiling in the organic phase has given a mixture of furanones 16 and 40 (Scheme 2.1). Heating of these mixtures with triethylamine resulted in 67% yield of furanone 16, total reaction duration being 14 h [77]. The process intermediates have not been studied, and the reaction mechanism has been explained using the scheme from [50].

Comparative study of oxidation of aldehyde 36 with H_2O_2 in water–1,2-dichloroethane and water–ethyl acetate two-phase systems in the presence of formic acid (1 : 0.8 molar ratio with respect to furfural) has been reported [75]. The yield of furanone 16 has been up to 60–62%, and the total yield of acids 22 and 47 has been 15–20%. Peroxyformic acid has been considered the oxidant in an organic phase. Alternatively, splitting of the cycle in a molecule of aldehyde 36 assisted by water in acidic medium has been suggested. This seems unlikely in view of stability of furan cycle in furfural owing to the conjugation with the carbonyl group.

The oxidation of furfural 36 with hydrogen peroxide in the presence of homogeneous acid catalysts in two-phase aqueous-organic systems has been performed [76]. The conditions favoring predominant formation of furanone 16 have been elaborated; it has been further hydrogenated in the presence of heterogeneous metal complex catalysts to afford γ-butyrolactone.

5-Methyl-2(5H)-furanone (53) is prepared by oxidation of 5-methylfurfural with 27-30% hydrogen peroxide at pH <4 and 60°C (yield 30%) [37] by a method similar to the production of 2(5H)-furanone [72].

A method for producing furanone (53) from 2-methylfuran is also described. Furanone 53 is synthesized by $KHWO_4$–catalyzed oxidation of 2–methylfuran with 5% solution of hydrogen peroxide in n–butanol (prepared from 30% aqueous H_2O_2). A molar ratio 2–methylfuran : H_2O_2 : catalyst = 1 : 2 : 0.05 is used [78]. A vigorously stirred reaction mixture is heated at 40–50°C until complete consumption of 2–methylfuran. The solvent is removed *in vacuo*, and the residue is treated with triethylamine (3% based on the starting 2–methylfuran) and refluxed at 70°C for 3 h. After removal of the catalyst, furanone 53 is isolated in 50% yield by vacuum distillation to collect fraction with b.p. 85—86°C (10 Torr).

Furanones 16 and 40 are the tautomeric forms of 2-hydroxyfuran (15) formed at an intermediate stage of the furfural oxidation by aqueous hydrogen peroxide (Scheme 2.1). The oxidation of 2-methylfurfural and 2-methylfuran formed the closest homologues of these furanones. The oxidation of hydroxyfuran 15 by hydrogen peroxide competes with the above tautomeric transformations. This reaction leads to the formation of hydroxyfuranone 7a or its closest homologue (if 5-methylfurfural used) (Scheme 2.1). The yield of furanone 7a does not exceed 4% in the absence of catalysts [60]. In this case, the tautomeric transformations of compound 15 to furanones 16 and 40 prevail.

Application of one of the following catalysts: vanadium (IV) or vanadium (V) compounds, molybdenum (VI) compounds, and SeO_2, results in the formation of their peroxide species with H_2O_2, which are significantly more strong oxidizers than hydrogen peroxide [60]. Under these conditions, hydroxyfuran 15 rapidly oxidizes into hydroxyfuranone 7a (Scheme 2.1) and this process predominates over the competing tautomeric transformations of enol 15 into lactones 16 and 40. Taking these results into account, we elaborated synthetic procedure to prepare 5-hydroxy-2(5H)-furanone (7a) [59, 79]. The highest yield of 7a (up to 90%) was achieved upon oxidation of furfural with 30% hydrogen peroxide in the presence of catalytic amounts of vanadium 2-oxonaphthenate (or other vanadium catalyst) and hydroquinone in a water–acetone mixed solvent at a molar ratio furfural : H_2O_2 : catalyst : hydroquinone = 1 : 3 : 0.003 : 0.0003. The reaction is carried out at 55–60 ^{0}C for 6 h, than the oxidate is neutralized, concentrated, and extracted with diethyl ether. Drying the organic layers with Na_2SO_4 and removal of the solvent afforded crystalline product 7a, which melts at 55–57 ^{0}C.

Other conditions to synthesize compound 7a by oxidation of furfural or furan with hydrogen peroxide give target product in lower yields.

The cyclic form of hydroxyfuranone 7a is contaminated with its open form 7b (Scheme 2.1). These forms exist in equilibrium, which is shifted towards the cyclic form at acidic pH (pH 0–4) approached during oxidation of furfural with hydrogen peroxide [48, 51].

The closest homologue of compound 7a, 5-hydroxy-5-methyl-2(5H)-furanone (36) was synthesized from 2-methylfuran. To a solution of 2-methylfuran in acetone, 30% aqueous H_2O_2, vanadyl acetylacetonate VO(acac)$_2$ as a catalyst, 18-crown-6, and hydroquinone are added. A molar ratio 2-methylfuran : H_2O_2 : VO(acac)$_2$: 18-crown-6 : hydroquinone is 1 :

3.2 : 2 : : 0.002 : 0.0007. Oxidation is carried out at 60°C for 6.5 h under vigorous stirring. Compound 36 was isolated in 75% yield. Compound 36 is a crystalline substance melting at 33–35°C.

Based on the domination of oxidative transformations of enol 15 into hydroxyfuranone 7a in the system furfural–H_2O_2–catalyst (SeO_2 or vanadium (IV,V) compounds) revealed, synthetic approaches towards derivatives 11, 72 and 73 are elaborated (Scheme 3.4). These procedures involve catalytic oxidation of furfural with aqueous hydrogen peroxide in homogeneous water-organic mixtures, in which organic solvent acts as a co-reagent.

5-Ethoxy-2(5*H*)-furanone (11) is synthesized by stirring furfural with a solution of SeO_2 in 5% aqueous hydrogen peroxide for 2 h at 50°C until complete conversion of aldehyde 36. The oxidate is concentrated under vacuum to remove 80% of water, the residue is diluted with ethanol and chloroform, and the mixture is refluxed with simultaneous azeotropic distillation of water until its complete removal. The residue is neutralized with concentrated aqueous $NaHCO_3$ solution, extracted with diethyl ether, and distilled *in vacuo* to give 70% yield of ethoxyfuranone 11 with b.p. 103—105°C (16 Torr). For the reaction, a molar ratio furfural: H_2O_2: catalyst: EtOH: chloroform = 1:3:0.05:2:1 is used.

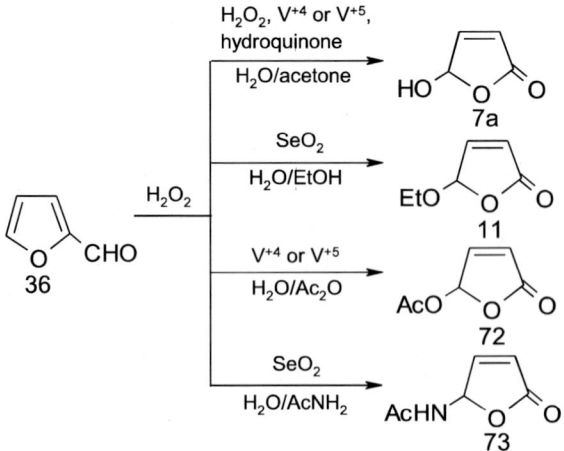

Scheme 3.4.

To synthesize 5-acetoxy-2(5*H*)-furanone (72), a stirred mixture of furfural, V_2O_5 or $VOSO_4$, acetone, and 30% aqueous hydrogen peroxide is heated at 50–60°C for 6 h. The oxidate is concentrated *in vacuo*, the residue is treated with acetic anhydride and heated at 60°C for 2 h. A molar ratio furfural : H_2O_2 : catalyst : acetone : acetic anhydride is 1 : 3 : 0.003 : 10 : 2. Compound 72 is isolated in 76% yield by vacuum distillation of the reaction mixture to collect a fraction with b.p. 116–120°C (5 Torr).

5-Acetamido-2(5*H*)-furanone (73) is synthesized as described above for 11 by oxidation of furfural (36) with 5% hydrogen peroxide in the presence of SeO_2 [59]. After removal of 80% water from the oxidate, the residue is treated with acetamide (molar ratio furfural : acetamide = 1 : 1.2) and the mixture is refluxed for 1 h. Compound 73 is isolated in 60% yield by column chromatography (elution with chloroform) as crystals melting at 115–117°C.

Thus, the finding of the synthetic conditions to access 5-hydroxy-2(5*H*)-furanone (7a) (Scheme 3.4) prompts the synthesis of its derivatives 11, 72 and 73 by catalytic oxidation of furfural without intermediate isolation of compound 7a from the oxidate. The earlier described syntheses of compounds 11, 72 and 73 started from furanone 7a, which was synthesized as a rule by photochemical oxidation of furfural [9].

Alkylated hydroxy derivatives of dihydrofurans were found among the products of furan and 2-methylfuran oxidation by hydrogen peroxide in the presence of $VOSO_4$ in aqueous-alcoholic media [83] (Section 1, Tables 1.1 and 1.2). Taking into account these data, a method for producing 2,5-dialkoxy-2,5-dihydrofurans 21a, 28 and 74-79 has been developed [84] (Scheme 3.5).

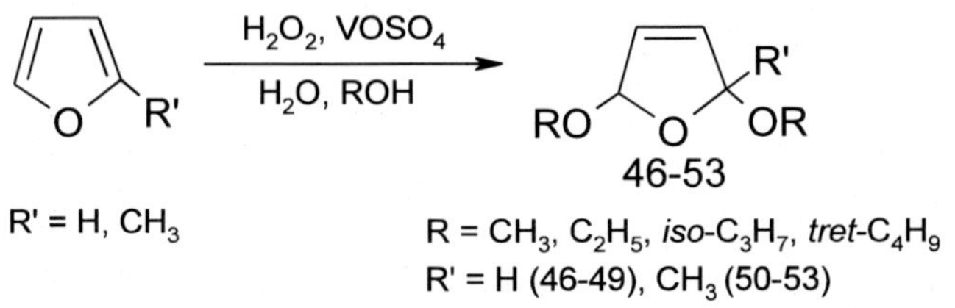

$$\text{furan-R'} \xrightarrow[H_2O, ROH]{H_2O_2, VOSO_4} \text{2,5-dialkoxy-dihydrofuran (46-53)}$$

R' = H, CH_3

R = CH_3, C_2H_5, *iso*-C_3H_7, *tret*-C_4H_9
R' = H (46-49), CH_3 (50-53)

Scheme 3.5.

To obtain dialkoxy-2,5-dihydrofurans 21a, 28 and 74-79 furan (for the synthesis of compounds 21a, 74-76) and 2-methylfuran (for the synthesis of compounds 28, 77-79) is oxidized with 30% aqueous hydrogen peroxide, diluted to the appropriate alcohol (methanol, ethanol, propan-2-ol, *tert.*–butanol) The process is carried out in the presence of $VOSO_4$ as a catalyst at 20-30^{0}C and intensive stirring. The molar ratio of furan compound, H_2O_2, alcohol and $VOSO_4$ is determined by the nature of furan and alcohol [84]. The oxidation products are extracted with sulfur ether, after removal of the solvent and distillation *in vacuo*, the compounds 21a, 28 и 74-79 are isolated in 25-38% yield depending on the nature of furan compound and the alcohol. The corresponding 5-alkoxy-2(5H)-furanones and 5-methyl-5-ethoxy-2(5H)-furanone are also isolated as by-products.

Thus, the differences in the reactions of furans and formylfurans with hydrogen peroxide and their causes have been considered in the sections 1-3 of the present review. The reactions of alkyl- and formylfurans with aqueous hydrogen peroxide have been shown to be sensitive to the effects of different reaction variables. By changing pH of the reaction medium, applying compounds of d-metals of V and VI groups or SeO_2 as catalysts, varying the reagent ratio and reaction temperature, and selecting the organic co-solvents soluble in water, it is possible to perform oxidation in homogeneous medium and to direct the reaction to obtain products of different types, namely, α-hydroxyhydroperoxides of furan aldehydes, 2-furoic acid and substituted 2-furoic acids, their esters and amides; 2(5*H*)-furanone, 5-hydroxy-2(5H)-furanones and their functional derivatives, as well as 2,5-dialkoxydihydrofurans and 5-alkoxy-2(5H)-furanones.

All these compounds are of great interest as the intermediates in chemical synthesis and biologically active substances. The elaborated synthetic procedures are advantageous over earlier described approaches.

REFERENCES

[1] Milas, N. A., and Walsh, W. L. (1935). Catalitic oxidation. I. Oxidation in furan series. *J. Amer. Chem. Soc.* 57 (8): 1389–1393.

[2] Lukevitz, E. Y., 1978. *Успехи химии фурана*. Рига: Знание. [*Advances in the chemistry of furan*. Riga: Zinatne].

[3] Shimanskaya, M. V., Giller, S. A., and Ioffe, I. I., 1971. *Парофазное контактное окисление фурановых соединений*. Рига: Зинатне. [*Vapor-phase contact oxidation of furan compounds*. Riga: Zinatne].

[4] Giller, S. A., and Shimanskaya, M. V., 1985. *К механизму контактных реакций окисления фурановых и пиридиновых соединений*. Москва: Наука. [*To the mechanism of contact oxidation reactions of furan compounds and pyridine compounds*. Moskow: Nauka].

[5] Shimanskaya, M. V., 1985. *Контактные реакции фурановых соединений*. Рига: Зинатне. [*Contact reactions of furan compounds*. Riga: Zinatne].

[6] Shimanskaya, M. V., 1990. Ванадиевые катализаторы окисления гетероциклических соединений. Рига: Зинатне. [*Vanadium catalysts for oxidation of heterocyclic compounds*. Riga: Zinatne].

[7] Mensah, T. (1986). Products identifed from photosensitised oxidation of selected Furanoid Feavor compounds. *J. Agr and Food Chem.*, 34 (2): 336–338.

[8] Schenk, G. O. (1953). Photochemische Reartionen. II. Uber die unsensibilisierte und photosensibilisierte autooxydation von Furanen. *Lieb. Ann. Chem.*, 584 (2/3): 156–176. [Photochemical reactions. II. About unsensitized and fotosensitized oxidation of furans. *Lieb. Ann. Chem*, 584 (2/3): 156–176].

[9] Schroeter, S. H., Appel, R., Brammer, R., and Schenk G. O. (1966). Maleinaldehydsaure und Fumaraldehydsaure. *Lieb. Ann. Chem.*, 697: 42–46. [Maleic. aldehydoacid and fumaric aldehydoacid. *Lieb. Ann. Chem.*, 697: 42–46.].

[10] Feringa, B. L., and Butselaar, R. I. (1982). A photooxidative analogue of Clauson-Kaas reaction. *Tetrahedron Lett.*, 23 (18): 1941–1942.

[11] Gollnick, K., and Griesbeck, A. (1985). Singlet oxygen photooxygenation of furans. Isolation and reactions of (4+2)-

cycloaddition products (unsaturated *sec.*–ozonides). *Tetrahedron*, 41 (11): 2057–2068.

[12] Fiedler, E., and Hacr, W. (1986). Elementary processes of O_2 ($^1\Delta g$) in gas phase. The chemical reaction with furan, 2-methylfuran and 2,5-dimetylfuran. *Oxid. Commun.*, 9 (3–4): 199–218.

[13] Masakatsu, M., Shigeru, N., Manabu, Hiroyuki, M., and Nobuko, W. (2000). Singlet oxygenation of 4-(4-*tert*-butyl-3,3-dimethyl-2,3-dihydrofuran-5-yl-2-pyridone: non-stereospecific 1,4-addition of singlet oxygen to a 1,3-diene system and thermal rearrangement of the resulting 1,4-endoperoxides to stable 1,2-dioxetanes. *Chem. Commun.*, 10: 821–822.

[14] Clauson-Kass, N., Limborg, F., and Farstrop, I. (1948). The alkoxylation of simple Furans and related reactions. *Acta Chemica Scandinavia*, 2: 109–115.

[15] Ogumi, Z., Ohhashi, S., and Tarehara Z. (1984). Methoxylation of furan on Pt-solid polimer electrolyte in the direct oxidation and bromine mediatory system. *J. Chem. Soc. Jap., Chem and Ind. Chem.*, 11: 1788–1793.

[16] Mederiou, M., and Montenegro, M. (1989). Microelectrode study of the methoxylation of furan. *Port. electrochim. acta*, 7: 47–50.

[17] Yoshida, K., and Fueno, I. (1971). Anodic oxidation. III. Controlled potential cyanomethoxylation of 2,5-dimethylfuran. *J. Org. Chem.*, 36 (11): 1523–1526.

[18] Nagiev, T. M. (1985). Сопряженные реакции окисления пероксидом водорода. *Успехи химии*, 44 (10): 1654-1673. [Conjugated reactions of oxidation with hydrogen peroxide. *Russian Chemical Reviews*, 44 (10): 1654-1673].

[19] Foot, S., and Wevler S. (1964). Olefin oxidation with excited singlet molecular oxygen. *J. Amer. Chem. Soc.*, 86 (18): 3879–3880.

[20] Milas N. A. (1927). Catalytic oxidation in aqueous solutions. I. The oxidation of furfural. *J. Amer. Chem. Soc*, 28 (8): 2005-2011.

[21] Milas N. A., Reeled R. l. and Magely O. L. (1954). Organic peroxides. XIX. α-hydroperohyethers and related peroxides. *J. Amer. Chem. Soc.*, 76 (9): 2322-2325.

[22] Salchinkin, A. P. (1959). Расщепление фурана горячим пергидролем под давлением. *Журнал прикладной химии*, 32 (8): 1605–1606. [Cleavage of the furan by hot perhydrol under pressure. *Russian Journal of Applied Chemistry*, 32 (8): 1605–1606].

[23] Gvozdetskaya, V. P., Kulnevich, V. G., and Lapkova, L. B. (1972). Получение диальдегидов и альдегидокисот при окислении фурана пергидролем. *Журнал прикладной химии*, 45 (2): 354-359. [Preparation of dialdehydes and aldehyde-acids by the oxidation of furan with perhydrol. *Russian Journal of Applied Chemistry,* 45 (2): 354-359].

[24] Gvozdetskaya, V. P., and Kulnevich, V. G., (1975). Изучение реакции фурана с перекисью водорода. Труды Краснодарского политехнического института. Химия и технология фурановых соединений, 66: 34-40. [*Proceedings of Krasnodar Polytechnic Institute. Chemistry and Technology of Furan Compounds*, 66: 34-40].

[25] Poskonin, V. V., Badovskaya, L. A., Gavrilova, S. P., and Kulnevich, V. G. (1989). Исследование в ряду замещенных бутан- и бутенолидов. V. Синтез 4-окси-2-бутен-4-олида и его производных на основе реакции фурфурола с перекисью водорода. Журнал органической химии, 25 (8): 1701–1705. [Study of a number of substituted butane and butenolides. V. Synthesis of 4-hydroxy-2-butene-4-olide and its derivatives by the reaction of furfural with hydrogen peroxide. *Russian Journal of Organic Chemistry,* 25 (8): 1701–1705].

[26] Poskonin, V. V., and Badovskaya, L. A. (1991). Реакции фурановых соединений с пероксидом водорода в присутствии ванадиевых катализаторов. *Химия гетероциклических соединений*, 11: 1462–1467. [Reactions of furan compounds with hydrogen peroxide in the presence of vanadium catalysts. *Chemistry of heterocyclic compounds*, 11: 1462–1467].

[27] Makarov, A. P., Gekhman, A. E., and Nekipelov, V. M. (1985). Комплекс ванадия(+5) с анион-радикалом дикислорода в катализе диспропорционирования пероксида водорода.

Известия Академии наук СССР. Серия химическая, 8: 1914–1917 [The complex of vanadium(+5) with the anion radical of dioxygen in catalysis of the hydrogen peroxide disproportionation. *Chemical Bulletin of the USSR, Chemical Series,* 8: 1914–1917].

[28] Brooks, H. B., and Sicilio, F. (1971). Electron spin resonance kinetic studies of the oxidation of vanadium (IV) by hydrogen peroxide. *Inorg. Chem.*, 10 (11): 2530–2534.

[29] Shiga, T., and Isomoto, A. (1969). Aromatic hydroxylation catalyzed by Fenton's reagents. An electron paramagnetic resonance study. I. Furans. *J. Phys. Chem.* 74 (4): 1139–1134.

[30] Maruthamuthu, P. (1985). Reaction of sulphate, phoshate and hydroxyl radicals with furan. An Electron spin resonance investigation in solution. *J. Chem. Soc., Faraday Trans.* 1, 81: 1979-1983.

[31] Poskonin, V. V., Badovskaya, L. A. Povarova, L. V. (1996). Реакции каталитического окисления фурановых и гидрофурановых соединений. 1. Общие закономерности окисления фурана в системе пероксид водорода – соединение ванадия(IV) в зависимости от типа растворителя и катализатора. *Химия гетероциклических соединений*, 5: 633–638. [Reactions of catalytic oxidation of furan and hydrofuran compounds. 1. General regularities of furan oxidation in the hydrogen peroxide-vanadium (IV) compound system depending on the type of solvent and catalyst. *Chemistry of heterocyclic compounds*, 5: 633–638].

[32] Poskonin, V. V., Badovskaya, L. A., and Povarova, L. V. (1998). Реакции каталитического окисления фурановых и гидрофурановых соединений. 3. Синтез 2,5-диэтокси-2,5-дигидрофурана в системе фуран – пероксид водорода – водный этанол – судьфат ванадила. *Химия гетероциклических соединений*, 7: 893–897. [Reactions of catalytic oxidation of furan and hydrofuran compounds. 3. Synthesis of 2,5-diethoxy-2,5-dihydrofuran in the system of furan – hydrogen peroxide – aqueous ethanol – vanadyl sulfate. *Chemistry of heterocyclic compounds,* 7: 893–897].

[33] Poskonin, V. V., Badovskaya, L. A., and Povarova, L. V. (1998). Реакции каталитического окисления фурановых и гидрофурановых соединений. 5. Гидрокси-, этоксидигидро фураны – новые продукты реакции фурана с пероксидом водорода. *Химия гетероциклических соединений*, 8: 1047–1054. [Reactions of catalytic oxidation of furan and hydrofuran compounds. 5. Hydroxy-, ethoxydihydrofurans as new products of the reaction of furan with hydrogen peroxide. *Chemistry of heterocyclic compounds*, 8: 1047–1054].

[34] Badovskaya, L. A., Poskonin, V. V., Povarova, L. V., and Ponomarenko, R. I. (1999). Реакции каталитического окисления фурановых и гидрофурановых соединений. 6. Синтетические возможности межфазного окисления фурана пероксидом водорода в присутствии соединений ванадия. *Химия гетероциклических соединений*, 10: 1322–1329. [Reactions of catalytic oxidation of furan and hydrofuran compounds. 6. Synthetic possibilities of interphase oxidation of furan by hydrogen peroxide in the presence of vanadium compounds. *Chemistry of heterocyclic compounds*, 10: 1322–1329].

[35] Poskonin, V. V., Badovskaya, L. A., and Povarova, L. V. (1999). Способ получения 2,5-диалкокси-2,5-дигидрофуранов. *Патент Российской федерации 2124508*. Опубликовано: Бюллетень изобретений, 1999 (1). [A method for the synthesis of 2,5-dialkoxy-2,5-dihydrofurans. *Russian patent 2124508*. Published: Bulletin of inventions, 1999 (1)].

[36] Gekhman, A. E., Moiseeva, N. I., and Moiseev, I. I. (1996). Исследование механизма реакции пероксида водорода с ионами ванадия. *Доклады Российской Академии наук*, 53: 349–353. [Study of the reaction mechanism of hydrogen peroxide with vanadium ions. *Doklady Chemistry*, 53: 349–353].

[37] Gekhman, A. E., Moiseeva, N. I., Minin, V. V., Larin, G. M., and Moiseev, I. I. (1997). Interaction between singlet dioxigen and superoxide anion radical coordinated with vanadium (V) ion. *Mendeleev. Comm.*, 6: 221–223.

[38] Badovskaya, L. A., and Povarova, L. V. (2009). Реакции окисления фуранов. *Химия гетероциклических соединений*, 9: 1283–1296. [Reactions of furans oxidation. *Chemistry of heterocyclic compounds*, 9: 1283–1296].

[39] Badovskaya, L. A., Povarova, L. V., and Kovalenko S. S. (2013). Превращения 2-метилфурана в системе пероксид водорода – ванадиевый катализатор – вода – этанол. *Вестник Казанского технологического университета*, 14: 93–96. [Transformations of 2-methylfuran in the hydrogen peroxide – vanadium catalyst – water – ethanol system. *Bulletin of Kazan Technological Univ.*, 14: 93–96].

[40] Povarova, L. V., Badovskaya, L. A., and Solovieva E. V. (2013). Влияние природы катализатора на окисление 2-метилфурана пероксидом водорода. *Современные наукоемкие технологии*, 9: 60–62. [The influence of the nature of the catalyst on the oxidation of 2-methylfuran with hydrogen peroxide. *Modern high technologies*, 9: 60–62].

[41] Badovskaya, L. A., and Kulnevich, V. G. (1969). Влияние условий реакции на характер продуктов окисления фурфурола перекисью водорода. *Химия гетероциклических соединений*, 2: 198-202. [The influence of reaction conditions on the nature of the oxidation products of furfural with hydrogen peroxide. *Chemistry of heterocyclic compounds*, 2: 198–202].

[42] Kulnevich, V. G., and Badovskaya, L. A. (1975). Реакции фурановых оксосоединений с перекисью водорода и надкислотами. *Успехи химии*, XLIV (7): 1256-1279. [Reactions of furan oxocompounds with hydrogen peroxide and superacids. *Russian Chemical Reviews*, XLIV (7): 1256-1279].

[43] Badovskaya, L. A. (1975). Избирательность реакций перекисного окисления фурановых альдегидов и их продукты. *Труды Краснодарского политехнического института. Химия и технология фурановых соединений*, 66: 1-18. [The selectivity of peroxide oxidation of furan aldehydes and their products. *Proceedings of Krasnodar Polytechnic Institute. Chemistry and Technology of Furan Compounds*, 66: 1–18].

[44] Badovskaya, L. A., Muzichenko, G. F., and Kulnevich, V. G. (1972). Способ получения оксигидроперекиси 5-нитрофурфурола. *Авторское свидетельство СССР 355169*. Опубликовано: Бюллетень изобретений, 1972 (31). [The method for 5-nitrofurfural oxyhydroperoxides preparation. *Copyright certificate of the USSR 355169*. Published: Bulletin of inventions, 1972 (31)].

[45] Badovskaya, L. A., Krapivin, G. D., Kaklyugina, T. Y., Kulnevich, V. G., and Muzichenko, G. F. (1975). Перекиси фурановых соединений. Синтез и некоторые свойства α-оксигидроперекисей фурановых альдегидов. *Журнал органической химии*, 1 (11): 2446–2447. [Peroxides of furan compounds. Synthesis and some properties of α-oxyhydroperoxides of furan aldehydes. *Russian Journal of Organic Chemistry*, 1 (11): 2446-2447].

[46] Kulnevich, V. G., Badovskaya, L. A., and Muzichenko, G. F. (1970). 2-Формилоксифуран – промежуточный продукт окисления фурфурола перекисью водорода. *Химия гетероциклических соединений*, 3: 582–584. [2-Formyloxyfuran as an intermediate of furfural oxidation with hydrogen peroxide. *Chemistry of heterocyclic compounds*, 3: 582–584].

[47] Badovskaya, L. A. (1978). Новая реакция получения низших оксогидрофуранов. *Химия гетероциклических соединений*, 10: 1314–1319. [A new reaction of lower oxohydrofurans preparation. *Chemistry of heterocyclic compounds*, 10: 1314–1319].

[48] Badovskaya, L. A., and Poskonin, V. V. (2018). Перегруппировки и таутомерные превращения гетероциклических соединений в гомогенных реакционных системах фурановый альдегид – H_2O_2 – растворитель. *Журнал общей химии*, 88 (8): 1245–1257. [Rearrangements and tautomeric transformations of heterocyclic compounds in homogeneous reaction systems furan aldehyde – H_2O_2 – solvent. *Russian Journal of General Chemistry*, 88 (8): 1245–1257].

[49] Badovskaya, L. A., Poskonin, V. V., and Ponomarenko, R. I. (2014). Влияние кислотно-основных свойств среды на реакции в системе фурановый альдегид – H_2O_2 – H_2O. *Журнал общей*

химии, 84 (6): 952–958. [Influence of acid-base properties of the medium on reactions in the system of furan aldehyde – H_2O_2 – H_2O. *Russian Journal of General Chemistry*, 84 (6): 952–958].

[50] Badovskaya, L. A., Latashko, V. M., Poskonin, V. V., Grunskaya, E. P., Tyukhteneva, Z. I., Rudakova, S. G., and Sarkisian, A. V. (2002). Реакции каталитического окисления фурановых и гидрофурановых соединений. 7. Получение 2(5H)-фуранона окислением фурфурола пероксидом водорода и некоторые его превращения. *Химия гетероциклических соединений*, 9: 1194–1203. [The reactions of catalytic oxidation of furan and hydrofuran compounds. 7. Preparation of 2(5H)-furanone by oxidation of furfural with hydrogen peroxide and some of its transformations. *Chemistry of heterocyclic compounds*, 9: 1194–1203].

[51] Badovskaya, L. A., and Poskonin, V. V. 2018. *Каталитические гомогенные и электрохимические реакции фурановых альдегидов с пероксидом водорода и синтезы функциональных производных фурана на их основе.* Краснодар: Кубанский государственный технологический университет. [*Catalytic homogeneous and electrochemical reactions of furan aldehydes with hydrogen peroxide and syntheses of furan functional derivatives on their basis.* Russia, Krasnodar: Kuban State Technological University].

[52] Grunskaya, E. P., Badovskaya, L. A., and Kaklyugina, T. Y. (2000). Кинетика и механизм образования фурановых пероксидов в реакции фурфурола с пероксидом водорода в присутствии и в отсутствие молибдата натрия. *Кинетика и катализ*, 41 (4): 495–498. [Kinetics and mechanism of furan peroxides formation in the reaction of furfural with hydrogen peroxide in the presence and absence of sodium molybdenum. *Kinetics and Catalysis*, 41 (4): 495–498].

[53] Muzichenko, G. F., Kaklyugina, T. Y., and Badovskaya, L. A. (1972). Изучение ассоциации в системах фурфурол – растворитель, фурфурол – перекись водорода. *Журнал физической химии*, 46 (9): 2391–2394. [Study of association in

furfural – solvent, and furfural – hydrogen peroxide systems. *Russian Journal of Physical Chemistry (A)*, 46 (9): 2391– 2394].

[54] Badovskaya, L. A., Kaklyugina, T. Y., and Kulnevich, V. G. (1977). Роль растворителя в реакции окисления фурфурола перекисью водорода. *Химия гетероциклических соединений*, 5: 603–607. [The role of the solvent in the oxidation of furfural with hydrogen peroxide. *Chemistry of heterocyclic compounds*, 5: 603–607].

[55] Badovskaya, L. A., Poskonin, V. V., and Povarova, L. V. (2017). Синтез функциональных производных фурана окислением фуранов и формилфуранов пероксидом водорода. Известия Российской академии наук. Серия химическая, 4: 593–599. [Synthesis of functional derivatives of furan by oxidation of furans and formylfurans with hydrogen peroxide. *Russian Chemical Bulletin*, 88 (8): 1245–1257].

[56] Krapivin, G. D., Yablonskiy, O. P., Chkalova, E. G., Badovskaya, L. A., and Kulnevich, V. G., (1970). Исследование комплексообразования в реакции окисления фурановых альдегидов пероксидом водорода. *Журнал физической химии*, 1 (6): 1399–1403. [Study of complex formation in the reaction of oxidation of furan aldehydes by hydrogen peroxide. *Russian Journal of Physical Chemistry (A)*, 1 (6): 1399–1403].

[57] Katritzki, Alan R., and Lagowski, Jeanne M., 1963. *Heterocyclic Chemistry*. London: Methuen & Co. Ltd., New York: John Wiley @ and Sons, Inc.

[58] March, Jerry, 1985. *Advanced Organic Chemistry. Reactions, Mechanisms and Structure*. New York: John Wiley @ and Sons, Inc.

[59] Poskonin, V. V., Badovskaya, L. A., Gavrilova, S. P., and Kulnevich, V. G. (1989). Исследование в ряду замещенных бутан- и бутенолидов. V. Синтез 4-окси-2-бутен-4-олида и его производных на основе реакции фурфурола с перекисью водорода. *Журнал органической химии*, 25 (8): 1701–1705. [Study of a number of substituted butane and butenolides. V. Synthesis of 4-hydroxy-2-butene-4-olide and its derivatives by the reaction of

furfural with hydrogen peroxide. *Russian Journal of Organic Chemistry*, 25 (8): 1701–1705].

[60] Badovskaya, L. A., and Poskonin, V. V. (2015). Влияние природы металла на каталитические реакции в системе фурфурол – H_2O_2 – H_2O – соль d-металла V или VI группы. *Кинетика и катализ*, 56 (2): 1-10. [The influence of the metal nature on the catalytic reaction in the system furfural – H_2O_2 – H_2O – salt of d-metal of V or VI group. *Kinetics and Catalysis*, 56 (2): 1-10].

[61] Strizhov, N. K., Poskonin, V. V., Badovskaya, L. A., and Kupina, E. P. (2002). Исследование в ряду замещенных бутан- и бутенолидов. XV. Превращения 4-гидрокси-2-бутен-4-олида в водных средах с различными значениями pH. *Журнал органической химии*, 38 (2): 273–277. [Study of a number of substituted butane and butenolides. XV. Transformations of 4-hydroxy-2-butene-4-olide in aqueous media with different pH values. *Russian Journal of Organic Chemistry*, 38 (2): 273–277].

[62] Poskonin, V. V., and Badovskaya, L. A. (2003). Необычное превращение 5-гидрокси-2(5H)-фуранона в водных растворах. *Химия гетероциклических соединений*, 5: 688–691. [Unusual conversion of 5-hydroxy-2(5H)-furanone in aqueous solutions. *Chemistry of heterocyclic compounds*, 5: 688–691].

[63] Gavrilova, S. P., Badovskaya, L. A., and Kulnevich, V. G. (1979). Кинетика и механизм образования 5-окси-2,5-дигидро ксифуранона-2 в реакции фурфурола с перекисью водорода в присутствии двуокиси селена. *Кинетика и катализ*, 20 (5): 1338–1343. [Kinetics and mechanism of 5-hydroxy-2,5-dihydroxyfuranone-2 formation in the reaction of furfural with hydrogen peroxide in the presence of selenium dioxide. *Kinetics and Catalysis*, 20 (5): 1338–1343].

[64] Visotskaya, N. A., Shevchuk, L. G., Gavrilova, S. P., Badovskaya, L. A., and Kulnevich, V. G. (1983). Радикальное гидроксилирование производных бензола и некоторых ароматических гетероциклов. *Украинский химический журнал*, 49 (8): 865–867. [Radical

hydroxylation of benzene derivatives and some aromatic heterocycles. *Ukranian Chemical Journal*, 49 (8): 865–867].
[65] Poskonin, V. V., Badovskaya, L. A., Grunskaya, E. P., Sarkisian, A. V., and Povarova, L. V. (1997). О влиянии растворителей в каталитических реакциях фурановых соединений с пероксидом водорода в присутствии оксидов металлов V и VI групп. В межвузовском сборнике: *Химия и технология фурановых соединений*, 57–65. Россия, Краснодар: Кубанский государст венный технологический университет. [On the effect of solvents in catalytic reactions of furan compounds with hydrogen peroxide in the presence of metal oxides of groups V and VI. In: *Chemistry and Technology of Furan Compounds*: 57-65. Russia, Krasnodar: Kuban State Technological University].
[66] Ponomarenko, R. I., Badovskaya, L. A., and Latashko, V. M. (2004). Способ получения фуранкарбоновой кислоты. *Патент Российской Федерации 2240999*. Опубликовано: Бюллетень изобретений, 2004 (33). [The method for furoic acid preparation. *Russian patent 2240999*. Published: Bulletin of inventions, 2004 (33)].
[67] Poskonin, V. V., Badovskaya, L. A., Sorotskaya, L. N., Povarova, L. V., Dedikova, T. G., Tlekhusezh, M. A., Kozhina, N. D., and Mitrofanova, S. P., 2015. Furans and formylfurans as the reagents for the syntheses of various classes of heterocyclic compounds. Paper presented at the *Int. Congr. on Heterocyclic Chem.* dedicated to 100 years anniversary of prof. Alexey Kost. Book of Abstr., Moskow, Russia, October 18-23.
[68] Krapivin, G. D., Badovskaya, L. A., and Kulnevich, V. G. (1977). Способ получения амидов фуранкарбоновой кислоты. *Авторское свидетельство СССР 530030*. Опубликовано: Бюллетень изобретений, 1977 (36). [The method for furoic acid amides preparation. *Copyright certificate of the USSR 530030*. Published: Bulletin of inventions, 1977 (36)].
[69] Poskonin, V. V., and Badovskaya, L. A. (1977). Способ получения амида фуран-2-карбоновой кислоты. *Авторское свидетельство*

СССР 1817456. Опубликовано: Бюллетень изобретений, 1990 (2). [The method for 2-furoic acid amide preparation. *Copyright certificate of the USSR 1817456* Published: Bulletin of inventions, 1990 (2)].

[70] Poskonin, V. V., and Badovskaya, L. A. (1994). Исследование в ряду замещенных бутан- и бутенолидов. IX. Синтезы новых 4-карбоксибутан-4-олидов и их производных. *Журнал органической химии*, 30 (7): 1001–1005. [Study of a number of substituted butane and butenolides. IX. Syntheses of novel 4-carboxybutane-4-olides and their derivatives. *Russian Journal of Organic Chemistry*, 30 (7): 1001–1005].

[71] Poskonin, V. V., and Badovskaya, L. A. (1998). 2,4-Дигидрокси-4-карбокси-3-Х-бутанолиды и способ их получения. *Патент Российской Федерации 2114837*. Опубликовано: Бюллетень изобретений, 1998 (19). [2,4-Dihydroxy-4-carboxy-3-X-butanolides and the method for their preparation. *Russian patent 2114837*. Published: Bulletin of inventions, 1998 (19)].

[72] Badovskaya, L. A., Muzichenko, G. F., Abramiants S. V., and Kulnevich, V. G. (1975). Способ получения кротонолактона. *Авторское свидетельство СССР 470516*. Опубликовано: Бюллетень изобретений, 1975 (18). [The method for crotonolactone preparation. *Copyright certificate of the USSR 470516*. Published: Bulletin of inventions, 1975 (18)].

[73] Cao, R., Lio, Ch., Lin, L. (1996). A convinient synthesis of 2(5H)-furanone. *Org. Prep. Proced. Int.*, 28(2): 215–219.

[74] Poskonin, V. V. (2009). Реакции каталитического окисления фурановых и гидрофурановых соединений. 9. Об особенностях и синтетических возможностях реакции фурфурола с водным пероксидом водорода в присутствии соединений ниобия (II) и (V). *Химия гетероциклических соединений*, 10: 1470–1477. [The reaction of catalytic oxidation and hydroturbomash furane compounds. 9. On the features and synthetic possibilities of furfural reaction with hydrogen peroxide in the presence of niobium (II) and

(V) compounds. *Chemistry of heterocyclic compounds*, 10: 1470–1477].

[75] Xiaodan, Li, Xiaocheng, Lan, and Tiefeng, Wang. Selective oxidation of furfural in a bi-phasic system with homogeneous acid catalyst (2016). *Catal. Today,* 276: 97–104.

[76] Xiaodan, Li, Xiaocheng, Lan, and Tiefeng, Wang. Highly selective catalytic conversion of furfural to γ-butyrolactone (2016). *Green Chem.,* 18: 638–642.

[77] Badovskaya, L. A., and Kulnevich, V. G. (1972). Способ получения ангеликалактона. *Авторское свидетельство СССР 348555.* Опубликовано: Бюллетень изобретений, 1972 (25). [The method for angelicalactone preparation. *Copyright certificate of the USSR 348555.* Published: Bulletin of inventions, 1972 (25)].

[78] Gavrilova, S. P., and Badovskaya, L. A. (1972). Способ получения 5-метил-2(5H)-фуранона. *Авторское свидетельство СССР 833965.* Опубликовано: Бюллетень изобретений, 1981 (20). [The method for 5-methyl-2(5H)-furanone preparation. *Copyright certificate of the USSR 833965.* Published: Bulletin of inventions, 1981 (20)].

[79] Poskonin, V. V., and Badovskaya, L. A. (1992). Способ получения 5-окси-2(5H)-фуранона. *Патент Российской Федерации 1715806.* Опубликовано: Бюллетень изобретений, 1992 (4). [The method for 5-hydroxy-2(5H)-furanone preparation. *Russian patent 1715806.* Published: Bulletin of inventions, 1992 (4)].

[80] Strizhov, N. K., Poskonin, V. V., Badovskaya, L. A., and Kupina, E. P. (2002). Исследование в ряду замещенных бутан- и бутенолидов. XV. Превращения 4-гидрокси-2-бутенолида в водных средах с различными значениями pH. *Журнал органической химии,* 38 (2): 273–277. [Study of a number of substituted butane and butenolides. XV. Transformations of 4-hydroxy-2-butenolide in aqueous media with different pH. *Russian Journal of Organic Chemistry,* 38 (2): 273–277].

[81] Badovskaya, L. A., and Gavrilova, S. P. (1979). Способ получения 2-оксо-5-этокси-2,5-дигидрофурана. *Авторское свидетельство*

СССР 651003. Опубликовано: Бюллетень изобретений, 1979 (9). [The method for 2-oxo-5-ethoxy-2(5H)-furanone preparation. *Copyright certificate of the USSR 651003*. Published: Bulletin of inventions, 1979 (9)].

[82] Poskonin, V. V., and Badovskaya, L. A. (1992). Способ получения 5-ацетокси-2(5Н)-фуранона. *Патент Российской Федерации 1703647*. Опубликовано: Бюллетень изобретений, 1992 (1). [The method for 5-acetoxy-2(5H)-furanone preparation. *Russian patent 1703647*. Published: Bulletin of inventions, 1992 (1)].

[83] Poskonin, V. V., Badovskaya, L. A., and Povarova, L. V. (1998). Реакции каталитического окисления фурановых и гидрофурановых соединений. 3. Синтез 2,5-диэтокси-2,5-дигидрофурана в системе фуран – пероксид водорода – водный этанол – сульфат ванадила. *Химия гетероциклических соединений*, 7: 893–897. [The reaction of catalytic oxidation of furan and hydrofuran compounds. 3. Synthesis of 2,5-diethoxy-2,5-dihydrofuran in the system furan – hydrogen peroxide – aqueous ethanol – vanadyl sulfate. *Chemistry of heterocyclic compounds*, 7: 893–897].

[84] Poskonin, V. V., and Badovskaya, L. A. (1999). Способ получения 2,5-диалкокси-2,5-дигидрофуранов. *Патент Российской Федерации 2124508*. Опубликовано: Бюллетень изобретений, 1999 (1). [The method for 2,5-dialcoxy-2,5-dihydrofurans preparation. *Russian patent 2124508*. Published: Bulletin of inventions, 1999 (1)].

BIOGRAPHICAL SKETCHES

Larisa A. Badovskaya

Affiliation: Kuban State Technological University, Dept. of Chemistry

Education: higher education; postgraduate studies

Research and Professional Experience:
Research area: Catalytic reactions of furan compounds with hydrogen peroxide; syntheses of biologically active substances based on these reactions and their products; the methodology for the synthesis of polyfunctional compounds from furans and hydrofuranones.

More than 350 scientific papers and patents.

PhD thesis (1967); doctoral dissertation (1983).

Gold medal of the 16th international salon of inventions "Archimedes-2013."

State grant "New methods of synthesis of hydrofuranones by the reaction of furfural with hydrogen peroxide in the presence of vanadium compounds," 1995–1996 (supported by the Russian Foundation for basic research).

State grant "Development of methods for the synthesis and fundamentals of technology for furanones and biologically active substances production," 2006–2007 (supported by the Russian Foundation for basic research and Administration of Krasnodar region).

Professional Appointments: Prof.

Honors: Honored scientist of the Russian Federation (1989), honorary worker of higher professional education of the Russian Federation (2000).

Order "Badge of Honor" (1971); medal "For valiant labor" (1970); breastplate "For excellent success in work in the field of higher education of the USSR" (1980); silver medal of VDNH "For success in national economy of the USSR" (1984); medal "For contribution to development of Kuban" III degree (1998); diploma of Legislative Assembly of Krasnodar Region (2003); diploma of Administration of Krasnodar Region (2008); medal "For outstanding contribution to development of Kuban" II degree (2018).

Publications from the Last 3 Years:

1. Badovskaya, L. A., and Poskonin, V. V. (2018). Перегруппировки и таутомерные превращения гетероциклических соединений в гомогенных реакционных системах фурановый альдегид – H_2O_2 – растворитель. *Журнал общей химии*, 88 (8): 1245–1257. [Rearrangements and tautomeric transformations of heterocyclic compounds in homogeneous reaction systems furan aldehyde – H_2O_2 – solvent. *Russian Journal of General Chemistry*, 88 (8): 1245–1257].
2. Badovskaya, L. A., Sorotskaya, L. N., Kozhina, N. D., Kaklugina, T. Y. (2018). Исследование в ряду замещенных бутан- и бутенолидов XVII. Замещенные 3-(фуран-2-илметилиден)фуран-2(3H)-оны и –дигидрофуран-2(3H)-оны. *Журнал органической химии,* 54 (7): 1027–1030.[Study in a number of substituted butane and butenolides. XVII. Substituted 3-(furan-2-ilmethyliden)furan-2(3H)-ones and –dihydrofuran-2(3H)-ones. *Russian Journal of Organic Chemistry*, 54 (7): 1027–1030].
3. Badovskaya, L. A., and Poskonin, V. V. 2018. *Каталитические гомогенные и электрохимические реакции фурановых альдегидов с пероксидом водорода и синтезы функциональных производных фурана на их основе*. Краснодар: Кубанский государственный технологический университет. [*Catalytic homogeneous and electrochemical reactions of furan aldehydes with hydrogen peroxide and syntheses of furan functional derivatives on their basis.* Russia, Krasnodar: Kuban State Technological University].
4. Badovskaya, L. A., Poskonin, V. V., and Povarova, L. V. (2017). Синтез функциональных производных фурана окислением фуранов и формилфуранов пероксидом водорода. Известия Российской академии наук. Серия химическая, 4: 593–599. [Synthesis of functional derivatives of furan by oxidation of furans and formylfurans with hydrogen peroxide. *Russian Chemical Bulletin*, 88 (8): 1245–1257].

5. Badovskaya, L. A., Tlekhusezh, M. A., and Nenko, N. I. (2017). Влияние гетарил-1,3-оксазолидинов на посевные качества семян озимой пшеницы. *Агрохимия*, 1: 46-49. [The influence of hetaryl-1,3-oxazolidine on sowing qualities of winter wheat seeds. *Agrochemistry,* 1: 46-49].
6. Badovskaya, L. A., Poskonin, V. V., and Povarova, L. V., 2016. Formation and transformations of tautomeric and isomeric forms of lower furanones in the reaction system «furfural – H_2O_2 – H_2O». Paper presented at the *XX Mendeleev Congr. on General and Applied Chemistry. Book of Abstr.*, Yekaterinburg, Russia, Ural branch of the Rus. Acad. of Sci.
7. Badovskaya, L. A., and Poskonin, V. V. (2015). Влияние природы металла на каталитические реакции в системе фурфурол – H_2O_2 – H_2O – соль d-металла V или VI группы. *Кинетика и катализ*, 56 (2): 1-10. [The influence of the metal nature on the catalytic reaction in the system furfural – H_2O_2 – H_2O – salt of d-metal of V or VI group. *Kinetics and Catalysis*, 56 (2): 1-10].
8. Poskonin, V. V., Badovskaya, L. A., Sorotskaya, L. N., Povarova, L. V., Dedikova, T. G., Tlekhusezh, M. A., Kozhina, N. D., and Mitrofanova, S. P., 2015. Furans and formylfurans as the reagents for the syntheses of various classes of heterocyclic compounds. Paper presented at the *Int. Congr. on Heterocyclic Chem.* dedicated to 100 years anniversary of prof. Alexey Kost. *Book of Abstr.*, Moskow, Russia, October 18-23.
9. Badovskaya, L. A., Dedikova, T. G., Povarova, L. V., and Nenko, N. I. (2015). Ростостимулирующая активность циклических фуран- и дигидрофурансодержащих ацеталей. *Агрохимия,* 6: 59-63. [Growth-stimulating activity of cyclic furan- and dihydrofuran-containing acetals. *Agrochemistry,* 6: 59-63].
10. Badovskaya, L. A., Povarova, L. V., and Nenko, N. I. (2015). Способ повышения посевных качеств семян яровой пшеницы. *Патент Российской Федерации 2561448*. Опубликовано: Бюллетень изобретений, 2015 (21). [Method of increasing sowing

qualities of spring wheat seeds. *Russian patent 2561448.* Published: Bulletin of inventions, 2015 (21)].

11. Povarova, L. V., Badovskaya, L. A., and Nenko, N. I. (2015). Способ повышения посевных качеств семян яровой пшеницы и устойчивости проростков к водному стрессу. *Патент Российской Федерации 2014115395/13.* Опубликовано: Бюллетень изобретений, 2015 (24). [Method for improving sowing qualities of spring wheat seeds and resistance of seedlings to water stress. *Russian patent 2014115395/13.* Published: Bulletin of inventions, 2015 (24)].

12. Tlekhusezh, M. A., Badovskaya, L. A., and Nenko, N. I. (2015). N-бензил-2-(3-бензил-2-тиофен-2-ил-1,3-оксазолидин-4-ил)ацет-амид, активирующий прорастание семян озимой пшеницы. *Патент Российской Федерации 2558139.* Опубликовано: Бюллетень изобретений, 2015 (21). [N-benzyl-2-(3-benzyl-2-thiophene-2-il-1,3-oxazolidin-4-il)acetamide, activating germination of winter wheat seeds. *Russian patent 2558139..* Published: Bulletin of inventions, 2015 (21)].

13. Badovskaya, L. A., Tlekhusezh, M. A., and Nenko, N. I. (2017). Влияние гетарил-1,3-оксазолидинов на посевные качества семян озимой пшеницы. *Агрохимия*, 1: 46-49. [The influence of hetaryl-1,3-oxazolidine on sowing qualities of winter wheat seeds. *Agrochemistry,* 1: 46-49].

Vladimir V. Poskonin

Affiliation: Kuban State Technological University, Dept. of Chemistry

Education: Higher education; doctorate

Research and Professional Experience:

Research area: Catalytic reactions of furan compounds with hydrogen peroxide; syntheses of biologically active substances based on these reactions and their products; the methodology for the synthesis of polyfunctional compounds from furans and hydrofuranones.
More than 200 scientific papers and patents.
PhD thesis (1990); doctoral dissertation (2002).
State grant "New methods of synthesis of hydrofuranones by the reaction of furfural with hydrogen peroxide in the presence of vanadium compounds," 1995–1996 (supported by the Russian Foundation for basic research).
State grant "Development of methods for the synthesis and fundamentals of technology for furanones and biologically active substances production," 2006–2007 (supported by the Russian Foundation for basic research and Administration of Krasnodar region).

Professional Appointments: Prof.

Honors:
Diploma of the Ministry of education of Russian Federation(2016).

Publications from the Last 3 Years:

1. Badovskaya, L. A., and Poskonin, V. V. (2018). Перегруппировки и таутомерные превращения гетероциклических соединений в гомогенных реакционных системах фурановый альдегид – H_2O_2 – растворитель. *Журнал общей химии*, 88 (8): 1245–1257. [Rearrangements and tautomeric transformations of heterocyclic compounds in homogeneous reaction systems furan aldehyde – H_2O_2 – solvent. *Russian Journal of General Chemistry*, 88 (8): 1245–1257].

2. Badovskaya, L. A., and Poskonin, V. V. 2018. *Каталитические гомогенные и электрохимические реакции фурановых альдегидов*

с пероксидом водорода и синтезы функциональных производных фурана на их основе. Краснодар: Кубанский государственный технологический университет. [*Catalytic homogeneous and electrochemical reactions of furan aldehydes with hydrogen peroxide and syntheses of furan functional derivatives on their basis*. Russia, Krasnodar: Kuban State Technological University].

3. Poskonin, V. V., and Yakovlev, V. V. (2018). Синтез малеиновой и яблочной кислот на основе электрохимической реакции фурфурола с пероксидом водорода. *Бутлеровские сообщения*, 54 (6): 56–61. [Synthesis of maleic and malic acids of furfural electrochemical reaction with hydrogen peroxide. *Butlerov Comm.*, 54 (6): 56–61].

4. Badovskaya, L. A., Poskonin, V. V., and Povarova, L. V. (2017). Синтез функциональных производных фурана окислением фуранов и формилфуранов пероксидом водорода. Известия Российской академии наук. Серия химическая, 4: 593–599. [Synthesis of functional derivatives of furan by oxidation of furans and formylfurans with hydrogen peroxide. *Russian Chemical Bulletin*, 88 (8): 1245–1257].

5. Poskonin, V. V., and Kovalenko, S. S., 2016. New directions of chemical neutralization of 5-hydroxymethylfurfural. Paper presented at the *Int. Scientific-practical Conf. Topical areas of fundamental and applied research X*. Proceedings of the Conference. North Charleston, SC, USA: Create Space.

6. Badovskaya, L. A., Poskonin, V. V., and Povarova, L. V., 2016. Formation and transformations of tautomeric and isomeric forms of lower furanones in the reaction system furfural – H_2O_2 – H_2O. Paper presented at the *20th Mendeleev Congr. on General and Applied Chemistry. Book of Abstr.*, Yekaterinburg, Russia, Ural branch of the Rus. Acad. of Sci.

7. Badovskaya, L. A., and Poskonin, V. V. (2015). Влияние природы металла на каталитические реакции в системе фурфурол – H_2O_2 – H_2O – соль d-металла V или VI группы. *Кинетика и катализ*, 56 (2): 1-10. [The influence of the metal nature on the catalytic

reaction in the system furfural – H_2O_2 – H_2O – salt of d-metal of V or VI group. *Kinetics and Catalysis*, 56 (2): 1-10].
8. Poskonin, V. V., Badovskaya, L. A., Sorotskaya, L. N., Povarova, L. V., Dedikova, T. G., Tlekhusezh, M. A., Kozhina, N. D., and Mitrofanova, S. P., 2015. Furans and formylfurans as the reagents for the syntheses of various classes of heterocyclic compounds. Paper presented at the *Int. Congr. on Heterocyclic Chem.* dedicated to 100 years anniversary of prof. Alexey Kost. *Book of Abstr.*, Moskow, Russia, October 18-23.
9. Poskonin, V. V., and Kovalenko, S. S. (2017). Способ повышения посевных качеств семян озимой пшеницы и устойчивости проростков к водному стрессу (засухе). *Патент Российской Федерации 2631690*. Опубликовано: Бюллетень изобретений, 2017 (27). [A method of increasing the sowing qualities of winter wheat seeds and seedlings resistance to water stress (drought). *Russian patent 2631690*. Published: Bulletin of inventions, 2017 (27)].

Larisa V. Povarova

Affiliation: Kuban State Technological University, Dept. of Chemistry

Education: higher education; postgraduate studies

Research and Professional Experience:
Research area: Catalytic reactions of furan compounds with hydrogen peroxide; syntheses of biologically active substances based on these reactions and their products; the methodology for the synthesis of polyfunctional compounds from furans and hydrofuranones.
More than 100 scientific papers and patents.
PhD thesis (1998).

Professional Appointments: Senior Lecturer of Faculty

Honors: State grant "Development of methods for the synthesis and fundamentals of technology for furanones and biologically active substances production," 2006–2007 (supported by the Russian Foundation for basic research and Administration of Krasnodar region).

Publications from the Last 3 Years:

1. Badovskaya, L. A., Poskonin, V. V., and Povarova, L. V. (2017). Синтез функциональных производных фурана окислением фуранов и формилфуранов пероксидом водорода. *Известия Российской академии наук. Серия химическая*, 4: 593–599. [Synthesis of functional derivatives of furan by oxidation of furans and formylfurans with hydrogen peroxide. *Russian Chemical Bulletin*, 88 (8): 1245–1257].
2. Badovskaya, L. A., Poskonin, V. V., and Povarova, L. V., 2016. Formation and transformations of tautomeric and isomeric forms of lower furanones in the reaction system furfural – H_2O_2 – H_2O. Paper presented at the *20th Mendeleev Congr. on General and Applied Chemistry. Book of Abstr.*, Yekaterinburg, Russia, Ural branch of the Rus. Acad. of Sci.
3. Poskonin, V. V., Badovskaya, L. A., Sorotskaya, L. N., Povarova, L. V., Dedikova, T. G., Tlekhusezh, M. A., Kozhina, N. D., and Mitrofanova, S. P., 2015. Furans and formylfurans as the reagents for the syntheses of various classes of heterocyclic compounds. Paper presented at the *Int. Congr. on Heterocyclic Chem.* dedicated to 100 years anniversary of prof. Alexey Kost. *Book of Abstr.*, Moskow, Russia, October 18-23.
4. Badovskaya, L. A., Dedikova, T. G., Povarova, L. V., and Nenko, N. I. (2015). Ростстимулирующая активность циклических фуран- и дигидрофурансодержащих ацеталей. *Агрохимия*, 6: 59-63. [Growth-stimulating activity of cyclic furan-and dihydrofuran-containing acetals. *Agrochemistry*, 6: 59-63].

5. Badovskaya, L. A., Povarova, L. V., and Nenko, N. I. (2015). Способ повышения посевных качеств семян яровой пшеницы. *Патент Российской Федерации 2561448.* Опубликовано: Бюллетень изобретений, 2015 (21). [Method of increasing sowing qualities of spring wheat seeds. *Russian patent 2561448.* Published: Bulletin of inventions, 2015 (21)].
6. Povarova, L. V., Badovskaya, L. A., and Nenko, N. I. (2015). Способ повышения посевных качеств семян яровой пшеницы и устойчивости проростков к водному стрессу. *Патент Российской Федерации 2014115395/13.* Опубликовано: Бюллетень изобретений, 2015 (24). [Method for improving sowing qualities of spring wheat seeds and resistance of seedlings to water stress. *Russian patent 2014115395/13.* Published: Bulletin of inventions, 2015 (24)].

In: Furan: Chemistry, Synthesis and Safety   ISBN: 978-1-53615-390-3
Editor: Ida Bailey   © 2019 Nova Science Publishers, Inc.

Chapter 2

EFFECTIVE SYNTHESIS OF NOVEL FURAN-FUSED PENTACYCLIC TRITERPENOIDS BY BASE-PROMOTED OR GOLD-CATALYZED 5-EXO DIG HETEROCYCLIZATION OF 2-ALKYNYL-3-OXOTRITERPENE ACIDS

A. Yu. Spivak* and R. R. Gubaidullin
Institute of Petrochemistry and Catalysis,
Russian Academy of Sciences, Ufa, Russia

ABSTRACT

The available plant metabolites, that are betulinic, ursolic and oleanolic acids, represent an important class of biologically active substances, which are in high demand in medicinal chemistry. These compounds are of interest for pharmacological investigations, as they

---

* Corresponding Author's E-mail: Spivak.ink@gmail.com.

exhibit various activities, such as anti-inflammatory, antiviral, hepatoprotective, antiparasitic and anticancer effects. Owing to the presence of easily transformable functional groups (OH, COOH, C=C), pentacyclic triterpenoids possess a good synthetic potential and are actively used as promising structural models for the discovery of new drugs. Currently, in order to increase the biological potential and bioavailability of native triterpene acids, their numerous synthetic analogues have been prepared. Among them, considerable attention has been given to heterocyclic triterpenoids with various nitrogen, sulfur, and oxygen heterocycles once they have been studied as antitumor, antiosteoporosis, anti-inflammatory and antileishmanial agents. Among this group of compounds, furan triterpenoid derivatives have not been reported in the literature. Meanwhile, polysubstituted furans represent an important class of oxygen heterocycles and occur as structural moieties in many natural products and pharmaceutically important substances. Furans are used in medicinal chemistry as useful intermediates in the synthetic transformations aimed at the development of new pharmaceutical agents. This review discloses the direct and atom-economical synthetic route to new [3,2-*b*]furan-fused pentacyclic triterpenoids via 5-exo-dig cyclization of accessible 2-alkynyl derivatives of betulonic, ursonic and oleanonic acids in the presence of the strong bases or catalyzed by transition gold complexes. Good prospects in these reactions have been revealed for highly alkynophilic gold-containing catalytic system $PPh_3AuCl/AgOTf$. Triterpenoids with a terminal or internal triple bond undergone cycloisomerization in the presence of the $PPh_3AuCl/AgOTf$ catalyst under very mild conditions. The generality of the method was demonstrated by the efficient preparation of furanfused triterpenoids, containing various functional groups, including hard Lewis base-sensitive groups such as CN, NO_2, or OAc, at the C-5' atom of the furan ring.

Keywords: pentacyclic triterpenoids, heterocycles, furans, alkynes, 5-exo-dig heterocyclization

INTRODUCTION

Polysubstituted furans represent an important class of oxygen-containing heterocycles, which occur as structural units in many biologically active compounds and pharmaceuticals (Banerjee, Hks, and Banerjee 2012). These five-membered heterocycles possess a high pharmacological potential and are actively used as useful intermediates in the synthetic and medicinal chemistry for the development of new drugs (Lee et al. 2005; Lipshutz 1986). Most of classical methods for the design of the furan ring are based on cyclocondensation of dicarbonyl compounds, the conditions of which are often inapplicable to compounds containing sensitive functional groups. Therefore, in the last decade, direct atom-economical synthetic routes to substituted furans based on intramolecular heterocyclization of acetylenic or allenic ketones and alcohols have been actively developed. These reactions are promoted by strong bases or acids (Arcadi et al. 1996; Vieser and Eberbach 1995; Vitale and Scilimati 2013) or catalyzed by transition metal complexes (Kirsch 2006; Brown 2005; Patil and Yamamoto 2007; Belting and Krause 2009; Mascareñas and López 2016; Rudolph and Hashmi 2012; Alonso, Beletskaya, and Yus 2004).

Our research group is engaged in systematic studies on the search for new biologically active compounds obtained by synthetic transformations of accessible secondary plant metabolites, that is, lupane, ursane, and oleane triterpenoids (Spivak et al. 2016; Gubaidullin et al. 2017; Spivak et al. 2017; Nedopekina et al. 2017; Spivak et al. 2014). Natural triterpenoids and their semisynthetic derivatives are of interest for pharmacological research, because they show various types of activities such as anti-inflammatory, hepatoprotective, antiviral, antiparasitic, and anticancer activities (Cichewicz and Kouzi 2004; Sarek et al. 2011; Mukherjee et al. 2006; Csuk 2014). Currently, in order to increase the biological potential and bioavailability of native triterpene acids, their numerous synthetic analogues have been prepared. Among them, considerable attention has been given to heterocyclic triterpenoids with various nitrogen, sulfur, and oxygen heterocyclic fragments at triterpenoid core (Kvasnica et al. 2015). The ketone carbonyl at C-3 of betulinic, ursolic and oleanolic acids was utilized in syntheses of various fused heterocycles at the 2,3-position of the triterpene skeleton including isoxazole, pyrazine, benzopyrazine, pyridine, indole and pyrazole rings (Laavola et al. 2016; Haavikko et al. 2014; Kumar et al. 2008; Urban et al. 2007; Xu et al. 2012; Wu et al. 2016). These triterpenoid derivatives modified with heterocyclic rings attached to the A-ring of the triterpene have shown anti-inflammatory (Laavola et al.

2016), antileishmanial (Haavikko et al. 2014), antitumor (Kumar et al. 2008; Urban et al. 2007) and antiosteoporosis (Xu et al. 2012; Wu et al. 2016) types of activities. Among this group of potentially biologically active compounds, furan triterpenoid derivatives are little-known compounds. Recently we developed an efficient method for the synthesis of 2-propargyl 3-oxo-triterpene acid derivatives of lupane-, ursane- and oleane-type based on α-alkylation with propargyl bromide of potassium enoxytriethylborates generated from 3-oxotriterpenes under the action of $KN(SiMe_3)_2$ – Et_3B (Spivak et al. 2016). The resulting triterpene compounds have been used in the synthesis of new furanotriterpenoids via base-promoted or gold-catalyzed 5-exo-dig heterocyclization of the pent-4-yn-1-one moiety in A-ring of the pentacyclic skeleton (Gubaidullin et al. 2017, 2018). A specific feature of this transformation is that the pent-4-yn-1-one moiety is incorporated in a polycyclic molecules, structurally related to steroids, whereas in many cases, acyclic alkynyl ketones have been used as the initial substrates for cyclization into furans (Arcadi et al. 1996; Vieser and Eberbach 1995; Vitale and Scilimati 2013; Kirsch 2006; Brown 2005; Patil and Yamamoto 2007; Belting and Krause 2009). To our knowledge, only one research group has described the synthesis of [3,2-b]furan-fused steroids through anionic annulation reaction of 4-pentynon moiety in the A-ring of steroid core (Arcadi and Rossi 1998).

EFFECTIVE SYNTHESIS OF NOVEL FURAN-FUSED PENTACYCLIC TRITERPENOIDS VIA ANIONIC 5-EXO DIG CYCLIZATION OF 2-ALKYNYL-3-OXOTRITERPENE ACIDS

The initial compounds 8-10 were synthesized by a reported method (Spivak et al. 2016) via the reaction of propargyl bromide with the enolate anion, which was formed by treating the methyl esters of betulonic 2, ursonic 4, and oleanonic acids 6 with $KN(SiMe_3)_2$ – Et_3B in 1,2–dimethoxyethane at room temperature. The reactions afforded C(2)-

Effective Synthesis of Novel Furan-Fused Pentacyclic ...

propargy triterpene acid derivatives 8-10 with equatorial-oriented α-propynyl groups (Scheme 1).

The cyclization conditions were selected in relation to lupane triterpenoid 8. Upon the reaction with superbases (Trofimov and Schmidt 2014) BuOK^t – DMSO, BuOK^t – DMF, BuOK^t – DME, KOH – THF, or KOH – DMSO, at room temperature over a period of 1–2 h, compound 8 was fully converted to a complex mixture of oligomeric compounds, in which the desired product 11a was not found. The use of KN(SiMe$_3$)$_2$ in DMSO gave heterocycle 11a in a yield not exceeding 36% (Scheme 2).

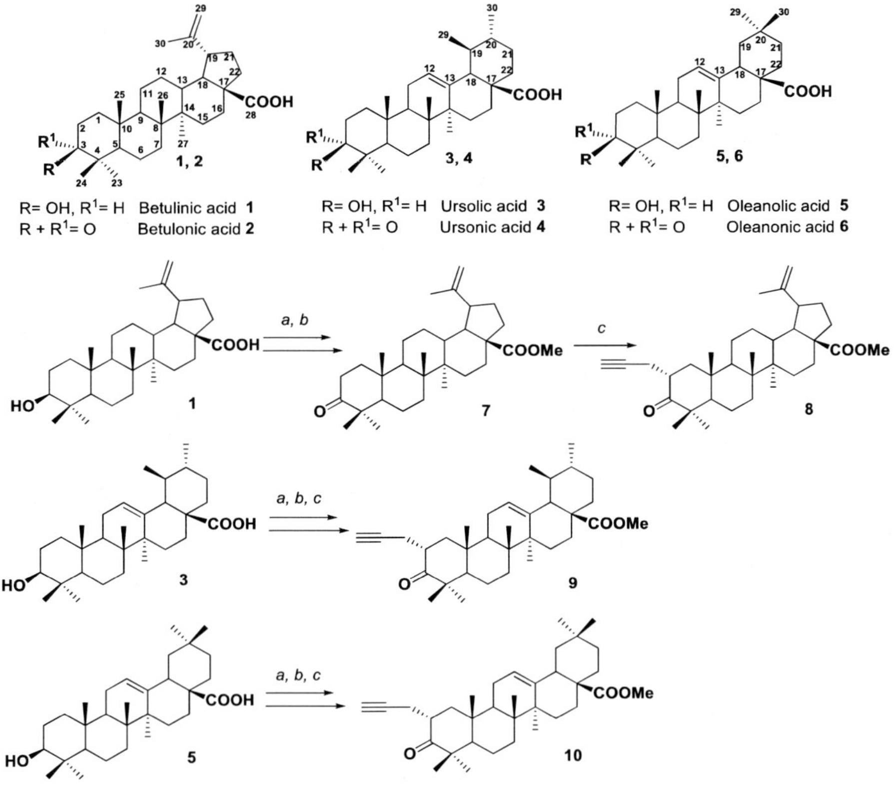

Reagents and conditions: a CrO$_3$, H$_2$SO$_4$, acetone or PCC, CH$_2$Cl$_2$. b CH$_2$N$_2$, Et$_2$O. c KN(SiMe$_3$)$_2$-Et$_3$B, DME.

Scheme 1. The preparation of C-2 propargyl triterpene acid derivatives 8-10.

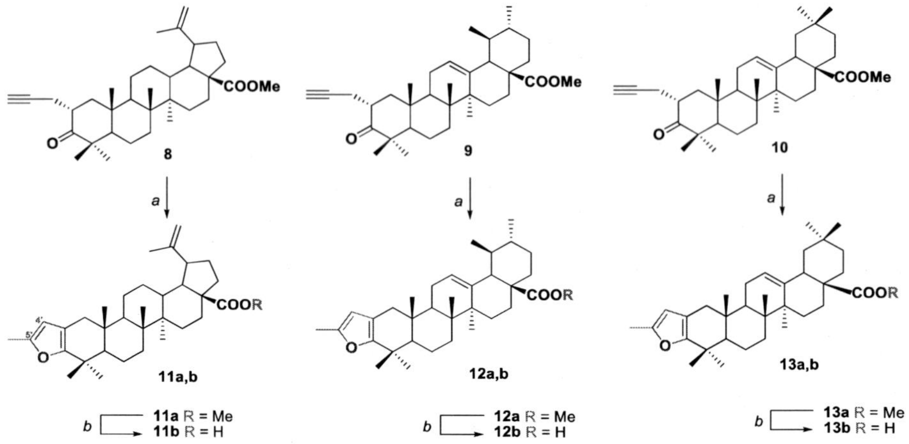

Reagents and conditions: *a* KN(SiMe$_3$)$_2$–Et$_3$B, DME, rt, Ar. *b* LiI, DMF, reflux, Ar.

Scheme 2. Synthesis of [3,2-*b*]furan fused triterpenoids 11a,b, 12a,b and 13a,b.

The expected compound 11a was obtained in a reasonable yield of 58% by treatment of terpenoid 8 with KN(SiMe$_3$)$_2$ in dimethoxyethane at room temperature for 30 min. At longer reaction times, the yield of furan derivative 11a was lower as a result of formation of oligomeric side products. Under the optimized conditions, methyl ursonate 9 and methyl oleonate 10 were converted to heterocyclic compounds 12a and 13a in 56% and 54% yields, respectively (Scheme 2, Table 1).

Demethylation of sterically hindered ester group in compounds 11a-13a via halogenolysis with LiI in DMF (Spivak et al. 2016) afforded 11b-13b in 54-56% yield (Scheme 2).

In order to broaden the applicability of this method, aryl groups with various substituents in the aromatic ring (4-Cl, 4-Br, 4-F, 3,4,5-OMe, 2-Me, 4-NO$_2$) were introduced into the terminal acetylenic moiety of compounds 8-10. These products were obtained in excellent yield (80-85%) by the Sonogashira reaction in the presence of PdCl$_2$(PPh$_3$)$_2$, CuI and Et$_3$N (Scheme 3).

The resulting alkynyl triterpenoid derivatives 14a-g, 17a, and 19a were successfully converted to triterpene furans 15a-g, 18a and 20a. Cleavage of methyl esters afforded compounds 16a-f, 18b and 20b. Triterpenoids with arylalkynyl substituents were more reactive towards this intramolecular cyclization than the substrates containing a terminal acetylenic bond. As opposed to cyclization of triterpenoid 8, compound 14a was transformed into furan derivative 15a in a good yield (54–73%) in the presence of various basic reagents: BuOK^t – DMSO, KN(SiMe$_3$)$_2$ – DMSO, BuOK^t – DME or KN(SiMe$_3$)$_2$ – DME. However the best yield of furan derivative 15a (73%) was obtained with the KN(SiMe$_3$)$_2$ – DME (Table 1). Among the testing 2-arylacetylenic derivatives of betulonic acid 14a-g only triterpenoid 14g (R = 4-NO$_2$C$_6$H$_4$) was problematic compound. Intramolecular cyclisation provided a 19% isolation yield of furanoterpenoid 15g and lead to the formation a large amounts of side-products.

Aryl-substituted acetylene derivatives 14a-g, 17a, and 19a were cycloisomerized on treatment with KN(SiMe$_3$)$_2$ – DME markedly faster (in 10-12 min) than propargyl-substituted triterpenoids 8-10 (30 min) to give the target reaction products 15a-g, 18a, and 20a in higher yields (70-77%).

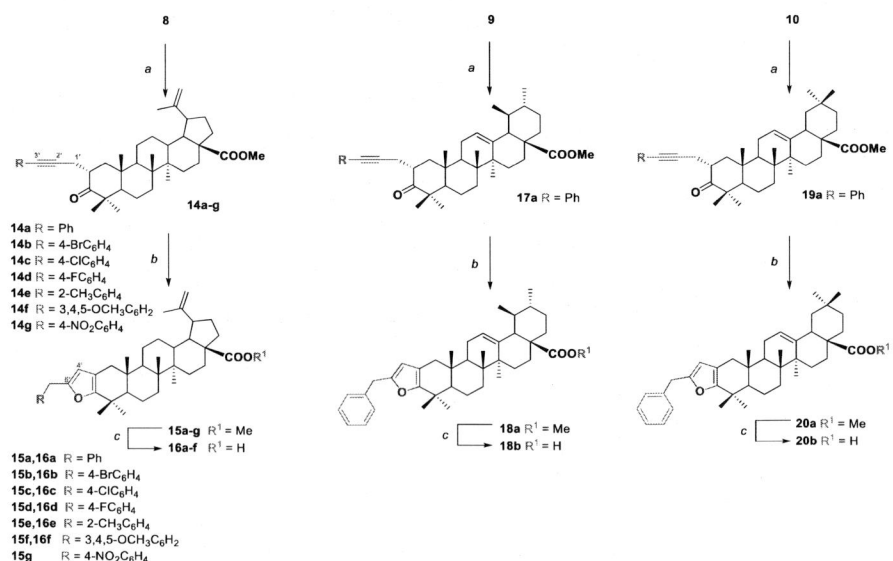

Reagents and conditions: a ArI, PdCl$_2$(PPh$_3$)$_2$, CuI, Et$_3$N, DMF, Ar, 20°C. b KN(SiMe$_3$)$_2$,DME, 20°C or BuOK^t, DME, 20°C. c LiI, DMF, reflux, Ar.

Scheme 3. Synthesis of [3,2-*b*]furan fused triterpenoids 15a-g, 16a-f, 18a,b and 20a,b.

Table 1. Reaction conditions for the synthesis of [3,2-*b*] furan-fused triterpenoids

Entry	C-2-alkynyl triterpenoids	R	Base	Solvent	[3,2-*b*]furan-fused triterpenoids	Yield[a] %
1	8	H	BuOK^t	DMSO	11a	0
2	8	H	BuOK^t	DMF	11a	0
3	8	H	BuOK^t	DME	11a	0
4	8	H	KOH	THF	11a	0
5	8	H	KOH	DMSO	11a	0
6	8	H	KN(SiMe$_3$)$_2$	DMSO	11a	36
7	8	H	KN(SiMe$_3$)$_2$	DME	11a	58
8	9	H	KN(SiMe$_3$)$_2$	DME	12a	56
9	10	H	KN(SiMe$_3$)$_2$	DME	13a	54
10	14a	Ph	BuOK^t	DMSO	15a	54
11	14a	Ph	BuOK^t	DME	15a	70
12	14a	Ph	KN(SiMe$_3$)$_2$	DMSO	15a	57
13	14a	Ph	KN(SiMe$_3$)$_2$	DME	15a	73
14	14b	4-BrC$_6$H$_4$	KN(SiMe$_3$)$_2$	DME	15b	71
15	14c	4-ClC$_6$H$_4$	KN(SiMe$_3$)$_2$	DME	15c	70
16	14d	4-FC$_6$H$_4$	KN(SiMe$_3$)$_2$	DME	15d	72
17	14e	2-CH$_3$C$_6$H$_4$	KN(SiMe$_3$)$_2$	DME	15e	77
18	14f	3,4,5-OCH$_3$C$_6$H$_2$	KN(SiMe$_3$)$_2$	DME	15f	73
19	14g	4-NO$_2$C$_6$H$_4$	KN(SiMe$_3$)$_2$	DME	15g	19
20	17a	Ph	KN(SiMe$_3$)$_2$	DME	18a	71
21	19a	Ph	KN(SiMe$_3$)$_2$	DME	20a	72

a. Isolated yield.

The structures of all new compounds were confirmed by conventional analytical methods. The ^{1}H and ^{13}C NMR spectra of furan-fused triterpenoids adequately reflected their structure. Indeed, the ^{13}C NMR spectrum of compound 11a exhibited no signals for the acetylene and carbonyl carbon atoms, indicating that these functional groups of the initial methyl betulonate 8 were transformed in the intramolecular cyclization.

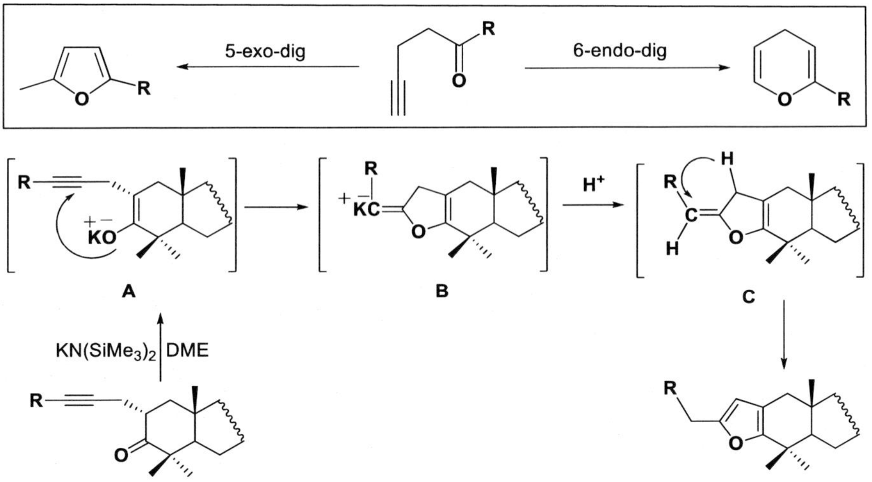

Figure 1. Assumed pathway to furan ring formation.

Apart from the characteristic signal for the quaternary C-20 carbon atom (150.57 ppm), the spectrum exhibited three new signals for quaternary carbon atoms (DEPT, HSQC) at 113.68, 149.57, and 154.44 ppm, which were assigned to C-2, C-5,' and C-3, respectively. The ^{1}H NMR spectrum contained, apart from the signals for protons at C-29, a new singlet for the vinylic H-4' proton at about 5.68 ppm. The proton signals for the methyl group at the furan C-5' atom occurred at 2.26 ppm. The (Me)C-5' carbon atom resonated at 13.71 ppm. The spectroscopic data indicated the presence of tetrasubstituted C-2–C-3 and trisubstituted C-4'−C-5' double bonds in compound 11a.

The base-promoted ring closure in acyclic alkynyl ketones and alcohols occurs via the addition of oxygen-based nucleophilic group to the carbon–carbon triple bond. Agreement with Baldwin rules for ring formation cyclization can proceed along two pathways (5-exo-dig or 6-endo-dig cyclization) to give either 2-alkylfurans or 4H-pyrans, respectively (Figure 1) (Baldwin 1976; Gilmore and Alabugin 2011). The KN(SiMe$_3$)$_2$ - promoted cyclization of compounds 8-10, 14a-g, 17a, and 19a proceeded with high regioselectivity as a 5-exo-dig cyclization according to the probable (Arcadi et al. 1996; Vieser and Eberbach 1995) pathway shown in Figure 1.

Apparently, elimination of the methine proton at the C-2 atom of ring A of triterpenoids 8-10, 14a-g, 17a, and 19a is followed by 5-exo-dig attack by the nucleophilic enolate oxygen on the triple bond of intermediate A to give intermediate B. The protodemetalation during hydrolysis of intermediate B affords unstable alkylidene dihydrofuran intermediate C, which undergoes a rapid isomerization to furan-fused triterpenoid.

Thus, an efficient synthetic route to biological interesting furan-fused pentacyclic triterpenoids with a furan moiety 2,3-annelated to the terpenoid skeleton has been developed. New heterocyclic triterpenoids of lupane-, ursane- and oleane- type have been obtained in moderate to good yields. However, under the conditions we developed, destruction of molecules containing functional groups sensitive to hard bases (OAc, NO_2, CN) took place. Therefore, in the continuation of our studies, we addressed the intramolecular heterocyclization of 2-alkynyl 3-oxo-triterpene acid derivatives catalyzed by metal complexes.

SYNTHESIS OF NOVEL [3,2-*B*]FURAN-FUSED PENTACYCLIC TRITERPENOIDS VIA GOLD - CATALYZED INTRAMOLECULAR HETEROCYCLIZATION OF 2-ALKYNYL-3-OXOTRITERPENE ACIDS

Currently, metal complex catalysis is widely used in the design of oxygen-containing heterocycles, including furans, based on acyclic acetylenic ketones and alcohols (Kirsch 2006; Brown 2005; Patil and Yamamoto 2007; Belting and Krause 2009; Mascareñas and López 2016; Rudolph and Hashmi 2012; Alonso, Beletskaya, and Yus 2004). Good prospects in these reactions have been revealed for gold-based catalytic systems (Belting and Krause 2009; Mascareñas and López 2016; Rudolph and Hashmi 2012). In many cases, gold complexes have been proved to be more efficient than the traditionally used Pd-, Ni-, or Rh-containing catalysts. The $Au(I)^+$ phosphine complexes are soft Lewis acids, which preferably activate soft electrophiles such as π-systems. The Au–alkyne complexes have shown excellent properties in nucleophilic addition reactions. The coordination of the [Au]$^+$ complex to an acetylenic bond considerably increases its electrophilicity (Mascareñas and López 2016; Rudolph and Hashmi 2012).

We studied the intramolecular heterocyclization of 2-alkynyl betulonic, ursonic, and oleanonic acid derivatives in the presence of the PPh$_3$AuCl/AgOTf catalyst (Gubaidullin et al. 2018). For comparison, the initial substrates were also involved in the anionic 5-exo-dig cyclization in the presence of KN(SiMe$_3$)$_2$ under conditions developed previously. It is noteworthy that this strategy of metal-involving synthesis of annelated furans rings has not been studied before for steroids and polycyclic steroid-like compounds.

Alkynyl triterpenoid derivatives 14a-n, 17a-e, and 19a-e were synthesized in high yields by the Sonogashira cross-coupling of compounds 8-10 with aryl iodides in the presence of PdCl$_2$(PPh$_3$)$_2$, CuI, and Et$_3$N (Scheme 4).

The heterocyclization of compounds 8-10, 14a-n, 17a-e, and 19a-e was performed in the presence of the highly alkynophilic PPh$_3$AuCl/AgOTf gold-containing system.

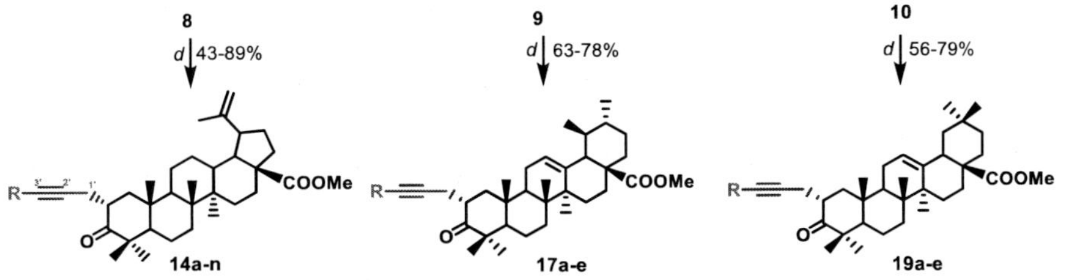

14a R = Ph	14i R = 4-F$_3$CC$_6$H$_4$	17a R = Ph	19a R = Ph
14b R = 4-BrC$_6$H$_4$	14j R = 2-pyridine	17b R = 2-thienyl	19b R = 2-thienyl
14c R = 4-ClC$_6$H$_4$	14k R = 2-thienyl	17c R = 2-acetylthienyl	19c R = 2-acetylthienyl
14d R = 4-FC$_6$H$_4$	14l R = 2-acetylthienyl	17d R = 4-NO$_2$C$_6$H$_4$	19d R = 4-NO$_2$C$_6$H$_4$
14f R = 3,4,5-OCH$_3$C$_6$H$_2$	14m R = 4-H$_2$NC$_6$H$_4$	17e R = 4-NCC$_6$H$_4$	19e R = 4-NCC$_6$H$_4$
14g R = 4-NO$_2$C$_6$H$_4$	14n R = 4-NCC$_6$H$_4$		
14h R = 4-Cl-2-BrC$_6$H$_4$			

Reagents and conditions: *a*, CrO$_3$, H$_2$SO$_4$, acetone, or PCC, CH$_2$Cl$_2$; *b*, CH$_2$N$_2$, Et$_2$O; *c*, KN(SiMe$_3$)$_2$-Et$_3$B, propargyl bromide, DME; *d*, ArI, PdCl$_2$(PPh$_3$)$_2$, CuI, Et$_3$N, DMF, Ar, 20 °C.

Scheme 4. Synthesis of C-2 alkynyl triterpenoid derivatives 14a-n, 17a-e and 19a-e.

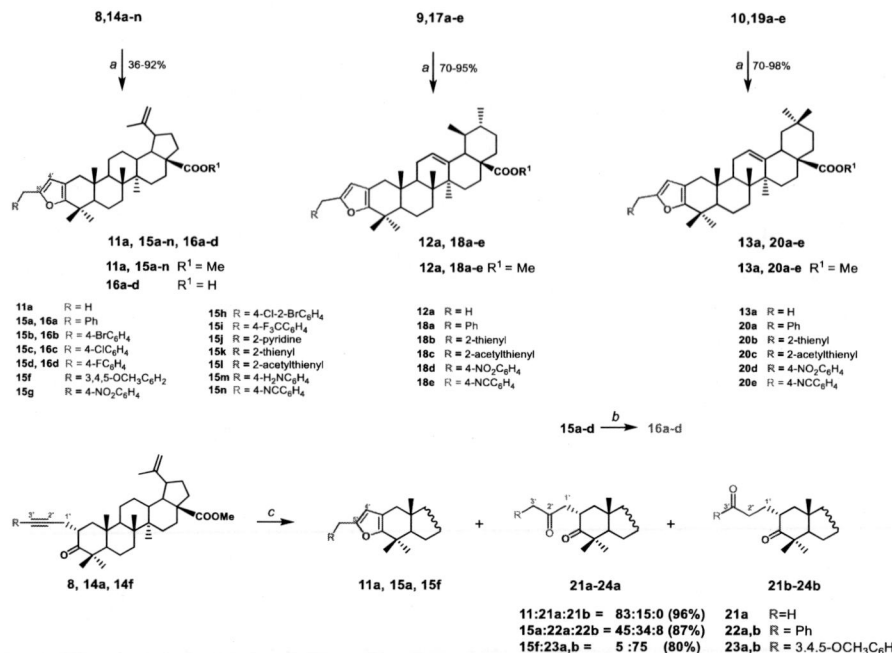

Scheme 5. Synthesis of [3,2-b]furan fused triterpenoids 11a-13a, 15a-n, 18a-e, 20a-e, and 16a-d in the presence of the PPh$_3$AuCl/AgOTf catalyst.

Previously, this cationic gold complex has shown a high catalytic activity towards the synthesis of substituted oxygen-containing heterocycles by cycloisomerization of simple acyclic acetylenic ketones and alcohols (Belting and Krause 2009).

The anionic 5-exo-dig cyclization was induced by KN(SiMe$_3$)$_2$ under conditions described in our previous publication (Gubaidullin et al. 2017). The metal-catalyzed heterocyclization of triterpenoids 8-10, 14a-n, 17a-e, and 19a-e occurred in the presence of 2 mol % PPh$_3$AuCl and 2 mol % AgOTf in toluene at room temperature and afforded furanoterpenoids 11a-13a, 15a-n, 21a-e, and 23a-e in high yields over a short period of time (Scheme 5, Table 2). The demethylation of the sterically hindered 28-ester group in the resulting furans on treatment with LiI in DMF afforded compounds 16a-d.

Table 2. Yields [3,2-*b*] of furan-fused triterpenoids in the anionic or Au-catalyzed 5-exo-dig cyclization

Entry	C-2-alkynyl triterpenoids	R	[3,2-*b*]furan-fused triterpenoids	Yield[a] % (Catalyst cyclization)	Yield[a] % (Anionic cyclization)
1	8	H	11a	92	53
2	9	H	12a	95	56
3	10	H	13a	98	54
4	14a	Ph	15a	72	73
5	14b	4-BrC$_6$H$_4$	15d	80	71
6	14c	4-ClC$_6$H$_4$	15c	79	70
7	14d	4-FC$_6$H$_4$	15d	73	72
8	14f	3,4,5-(CH$_3$O)$_3$C$_6$H$_2$	15f	36	73
9	14g	4-O$_2$NC$_6$H$_4$	15g	95	19
10	14h	4-Cl-2BrC$_6$H$_2$	15h	76	-[b]
11	14i	4-F$_3$CC$_6$H$_4$	15i	82	-[b]
12	14j	2-pyridine	15j	-[b]	44
13	14k	2-thienyl	15k	81	38
14	14l	2-acetylthienyl	15l	83	11
15	14m	4-H$_2$NC$_6$H$_4$	15m	-[b]	-[b]
16	14n	4-NCC$_6$H$_4$	15n	70	-[b]
17	17a	Ph	18a	73	71
18	17b	2-thienyl	18b	80	48
19	17c	2-acetylthienyl	18c	74	8
20	17d	4-O$_2$NC$_6$H$_4$	18d	91	12
21	17e	4-NCC$_6$H$_4$	18e	76	-[b]
22	19a	Ph	20a	70	72
23	19b	2-thienyl	20b	75	44
24	19c	2-acetylthienyl	20c	78	9
25	19d	4-O$_2$NC$_6$H$_4$	20d	80	8
26	19e	4-NCC$_6$H$_4$	20e	90	-[b]

a. Isolated yield; b. Traces.

The results summarized in Table 2 indicate that catalysis by metal complexes in the 5-exo-dig cyclization of the compounds in question provides considerable advantages over the anionic cyclization. Indeed, with this catalyst, the cyclization of 2-propargyl betulonic, ursonic, and oleanonic acids 8-10 with the terminal alkyne moiety proceeded in 5-10 min at room temperature to give target furanoterpenoids in quantitative yields (92-98%, entries 1-3), which were much higher than the yields of these products in the anionic 5-exo-dig cyclization (53-56%). The conditions of the catalytic reaction were tolerable for compounds containing aryl or hetaryl groups, such as p-$NO_2C_6H_4$, p-NCC_6H_4, 2-thienyl, and 5-acetyl-2-thienyl, at the alkyne moiety (entries 9, 13, 14, 16, 18-21, 23-26). The 5-exo-dig cyclization of these compounds induced by $KN(SiMe_3)_2$ was accompanied by destruction of the reaction products and, hence, the desired furanotriterpenoids were not obtained or obtained in only 8-48% yields. Meanwhile, the gold complex-catalyzed heterocyclization of triterpenoids 14b-d, 14h, 14i had low sensitivity to the nature of electron-deficient aryl groups (Table 2). The duration of reaction for compounds 14b-d, 14h, 14i did not exceed 1.5 h and the yields of target furanotriterpenoids 15b-d, 15h, 15i were 73-82%. In contrast to cyclization of triterpenoids 14b-d, 14h, 14i, cyclization of substrate 13h containing the electron-enriched aryl group, 3,4,5-$(OMe)_3C_6H_2$, at the acetylenic moiety proceeded over a longer period of time (4 h) and gave furan 15f in 36% yield. We suggest that in this case, the electron-donating substituent has a negative effect on the electrophilicity of the C≡C bond; meanwhile, the electrophilic activation of the acetylenic moiety is the driving force for triggering the initial step of the 5-exo-dig reaction. The attempts to accomplish the metal-catalyzed cyclization of compound 14j or 14m containing pyridine or aniline groups resulted in a complex mixture of products. Probably, successful heterocyclization of compounds 14j and 14m both with catalysis by metal complexes and in the presence of a hard base requires the use of protecting groups.

In all experiments, when traces of water were present in the reaction mixture, the competing $PPh_3AuCl/AgOTf$-catalyzed hydration of alkynes took place, resulting in the corresponding diketones. The content of these side products in the reaction mixture considerably increased for substrates that were less prone to cyclization. It is worth noting that the gold-catalyzed hydration and hydroalkoxylation belong to a thoroughly described class of reactions, which are widely used in organic synthesis, including targeted synthesis of natural products and pharmaceuticals (Goodwin and Aponick 2015; Arcadi et al. 2009; Hintermann and Labonne 2007; Leyva and Corma 2009).

The hydration of terminal alkynes proceeds with a high degree of selectivity giving Markovnikov products, but the hydration of internal alkynes gives, most often, regioisomers (Goodwin and Aponick 2015). In our control experiments, the PPh$_3$AuCl/AgOTf-induced transformation of propargylbetulonic acid 8 in toluene containing traces of H$_2$O proceeded regiospecifically at room temperature to give a mixture of furan 11a and 1,4-diketone 21a in 83:15 ratio (the ratio of isolated products). The compounds were isolated in a pure state in a total yield of 96% by chromatography of the mixture on a SiO$_2$ column. The reaction of terpenoid 14a under the same conditions afforded a mixture of diketone regioisomers 22ab and furanoterpenoid 15a (the 15a:22a:22b product ratio was 45:34:8 after the separation by column chromatography on SiO$_2$) (Scheme 5). The transformation of terpenoid 14f, which is less prone to cycloisomerization, in non-dried toluene afforded diketones 23ab as the major products. The content of furan 15f in the mixture did not exceed 5% in this case. Furanoterpenoid 15f and 1,4-diketone 23a were isolated as narrow chromatographic fractions from a mixture of compounds difficult to separate, using column chromatography on SiO$_2$.

In control experiments with terpenoids 8 and 14a, we found that AgOTf or the Ph$_3$PAuCl complex without addition of AgOTf do not initiate the cycloisomerization. It is known that chlorine containing Au(I) compounds are activated by silver salts via replacement of a chloride anion by the anion of a silver salt (Zhdanko and Maier 2015). Therefore, our reaction was apparently initiated by the Ph$_3$PAuOTf active complex formed *in situ*. In recent years, mixed Ag-Au bimetallic intermediates, which may also account for the silver effect, have been identified in the studies concerned with the mechanisms of Au(I)-catalyzed reactions (Zhdanko and Maier 2015).

Basing on the commonly accepted views of gold complex-catalyzed cycloisomerization of acetylenic or allenic ketones and alcohols (Belting and Krause 2009; Mascareñas and López 2016), we present a possible pathway to furanotriterpenoids (Figure 2). The Au(I)$^+$phosphonium complexes are soft Lewis acids, which preferably activate soft electrophiles such as π-systems. Therefore, the catalytic cycle is initiated by coordination of the Au(I) catalyst to the triple bond in the alkynyl moiety of the terpenoid. The gold cationic complex A thus formed promotes the 5-exo-dig attack of the nucleophilic oxygen atom on the activated C≡C bond with simultaneous elimination of the H-C(2) proton in triterpenoid ring A. This furnishes gold-containing zwitter-ion intermediate B. The protodeauration of intermediate B leads to the release of the catalyst and formation of unstable alkylidenefuran triterpenoid derivative C, which undergoes fast isomerization to be converted to the reaction product.

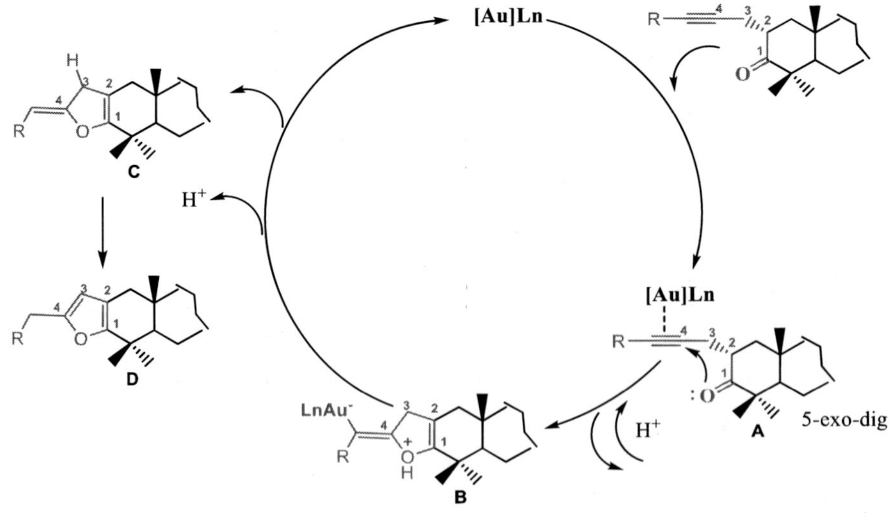

Figure 2. Presumptive pathway to [3,2-*b*]furan-substituted triterpenoids.

Previously, Pd-catalyzed rearrangements of acetylenic ketones to substituted furans have been reported in the literature. However, these methods were mainly limited to annelation of the furan ring to arenes (Omelchuk, Tikhomirov, and Shchekotikhin 2016).

A general effective method for the synthesis of substituted furans via Cu-catalyzed cycloisomerization of acyclic α,β-alkynyl ketones has been developed (Kel'i and Gevorgyan 2002). Using substrates 8 and 14a, we attempted to carry out the cycloisomerization of 2-alkynyl triterpenoid derivatives under the action of $PdCl_2(Ph_3P)_2$, $Pd(OAc)_2$, or $Pd(OAc)_2$ with triphenylphosphine added. The reactions were carried out in DME, THF, or toluene at room temperature or at 50-100°C. However, our attempts did not meet with success. The alkynyl triterpenoids either did not react under these conditions or produced unidentified product mixtures. The Cu(I) catalysis did not initiate cycloisomerization of triterpenoids 8 and 14a under the reported conditions either (Kel'i and Gevorgyan 2002).

CONCLUSION

In this study we have developed an efficient synthetic route to biological interesting furan-fused pentacyclic triterpenoids with a furan moiety 2,3-annelated to the terpenoid skeleton. New furan-fused triterpenoids have been obtained by base-promoted or gold-catalyzed ($PPh_3AuCl/AgOTf$) cycloisomerization of the pent-4-yn-1-one moiety in ring A of the 2-alkynyl-3-oxotriterpene acids of lupane-, ursane- and oleane- type. Triterpenoids containing terminal as well as internal triple bond were cyclized in high to excellent yields under mild reaction conditions. The protocols exhibited high regioselectivity, favoring the 5-exo-dig products. The generality of the gold-catalyzed reaction was demonstrated by the efficient preparation of furan-fused triterpenoids containing various functional groups, including hard Lewis base-sensitive groups such as CN, NO_2, or OAc.

REFERENCES

Alonso, Francisco, Irina P. Beletskaya, and Miguel Yus. 2004. "Transition-Metal-Catalyzed Addition of Heteroatom-Hydrogen Bonds to Alkynes." *Chemical Reviews* 104 (6): 3079–3159. https://doi.org/10.1021/cr0201068.

Arcadi, Antonio, Maria Alfonsi, Marco Chiarini, and Fabio Marinelli. 2009. "Sequential Gold-Catalyzed Reactions of 1-Phenylprop-2-Yn-1-Ol with 1,3-Dicarbonyl Compounds." *Journal of Organometallic Chemistry* 694 (4): 576–82. https://doi.org/10.1016/j.jorganchem.2008.12.013.

Arcadi, Antonio, Fabio Marinelli, Elena Pini, and Elisabetta Rossi. 1996. "Base Promoted Reactions of 4-Pentynones." *Tetrahedron Letters* 37 (19):3387–90. https://doi.org/10.1016/0040-4039(96)00553-9.

Arcadi, Antonio, and Elisabetta Rossi. 1998. "Synthesis of Functionalised Furans and Pyrroles through Annulation Reactions of 4-Pentynones." *Tetrahedron* 54 (50): 15253–72. https://doi.org/10.1016/S0040-4020(98)00953-3.

Baldwin, Jack E. 1976. "Rules for Ring Closure." *Journal of the Chemical Society, Chemical Communications* 0 (18): 734–36. https://doi.org/10.1039/c39760000734.

Banerjee, Rumpa, Kumar Hks, and Mrityunjay Banerjee. 2012. "Medicinal Significance of Furan Derivatives : A Review." *International Journal of Review in Life Sciences* 2 (1):7–16. https://ru.scribd.com/document/325738489/published-pdf-078-6-02-10052.

Belting, Volker, and Norbert Krause. 2009. "Gold-Catalyzed Cycloisomerization of Alk-4-Yn-1-Ones." *Organic & Biomolecular Chemistry* 7 (6):1221. https://doi.org/10.1039/b819704k.

Brown, Richard C D. 2005. "Developments in Furan Syntheses." *Angewandte Chemie International Edition* 44 (6):850–52. https://doi.org/10.1002/anie.200461668.

Cichewicz, Robert H., and Samir A. Kouzi. 2004. "Chemistry, Biological Activity, and Chemotherapeutic Potential of Betulinic Acid for the

Prevention and Treatment of Cancer and HIV Infection." *Medicinal Research Reviews* 24 (1):90–114. https://doi.org/10.1002/med.10053.

Csuk, René. 2014. "Betulinic Acid and Its Derivatives: A Patent Review (2008 – 2013)." *Expert Opinion on Therapeutic Patents* 24 (8):913–23. https://doi.org/10.1517/13543776.2014.927441.

Gilmore, Kerry, and Igor V Alabugin. 2011. "Cyclizations of Alkynes: Revisiting Baldwin's Rules for Ring Closure." *Chemical Reviews* 111 (11): 6513–56. https://doi.org/10.1021/cr200164y.

Goodwin, Justin a., and Aaron Aponick. 2015. "Regioselectivity in the Au-Catalyzed Hydration and Hydroalkoxylation of Alkynes." *Chemical Communications* 51 (42): 8730–41. https://doi.org/10.1039/C5CC00120J.

Gubaidullin, Rinat R., Rezeda R. Khalitova, Darya A. Nedopekina, and Anna Yu. Spivak. 2018. "Homo- and Cross Coupling of C-2 Propargyl Substituted Triterpenoic Acids: Synthesis of Novel Symmetrical and Unsymmetrical Triterpene 1,3-Diynes." *Chemistry Select* 3 (47):13526–29. https://doi.org/10.1002/slct.201803522.

Gubaidullin, Rinat R., Darina S. Yarmukhametova, Darya A. Nedopekina, Rezeda R. Khalitova, and Anna Yu Spivak. 2017. "Effective Synthesis of Novel Furan-Fused Pentacyclic Triterpenoids via Anionic 5-Exo Dig Cyclization of 2-Alkynyl-3-Oxotriterpene Acids." *Arkivoc* 2017 (5):100–116. https://doi.org/10.3998/ark.5550190.p010.142.

Haavikko, Raisa, Abedelmajeed Nasereddin, Nina Sacerdoti-Sierra, Dmitry Kopelyanskiy, Sami Alakurtti, Mari Tikka, Charles L. Jaffe, and Jari Yli-Kauhaluoma. 2014. "Heterocycle-Fused Lupane Triterpenoids Inhibit Leishmania Donovani Amastigotes." *Med Chem Comm* 5 (4):445–51. https://doi.org/10.1039/c3md00282a.

Hintermann, Lukas, and Aurélie Labonne. 2007. "Catalytic Hydration of Alkynes and Its Application in Synthesis." *Synthesis* 2007 (8):1121–50. https://doi.org/10.1055/s-2007-966002.

Kel'i, Alexander V., and Vladimir Gevorgyan. 2002. "Efficient Synthesis of 2-Mono- and 2,5-Disubstituted Furans via the CuI-Catalyzed Cycloisomerization of Alkynyl Ketones †." *The Journal of Organic Chemistry* 67 (1):95–98. https://doi.org/10.1021/jo010832v.

Kirsch, Stefan F. 2006. "Syntheses of Polysubstituted Furans: Recent Developments." *Organic & Biomolecular Chemistry* 4 (11):2076. https://doi.org/10.1039/b602596j.

Kumar, Vivek, Nidhi Rani, Pawan Aggarwal, Vinod K. Sanna, Anu T. Singh, Manu Jaggi, Narendra Joshi, Pramod K. Sharma, Raghuveer Irchhaiya, and Anand C. Burman. 2008. "Synthesis and Cytotoxic Activity of Heterocyclic Ring-Substituted Betulinic Acid Derivatives." *Bioorganic and Medicinal Chemistry Letters* 18 (18):5058–62. https://doi.org/10.1016/j.bmcl.2008.08.003.

Kvasnica, Miroslav, Milan Urban, Niall J. Dickinson, and Jan Sarek. 2015. "Pentacyclic Triterpenoids with Nitrogen- and Sulfur-Containing Heterocycles: Synthesis and Medicinal Significance." *Natural Product Reports* 32 (9). Royal Society of Chemistry:1303–30. https://doi.org/10.1039/C5NP00015G.

Laavola, Mirka, Raisa Haavikko, Mari Hämäläinen, Tiina Leppänen, Riina Nieminen, Sami Alakurtti, Vaînia M. Moreira, Jari Yli-Kauhaluoma, and Eeva Moilanen. 2016. "Betulin Derivatives Effectively Suppress Inflammation in Vitro and in Vivo." *Journal of Natural Products* 79 (2):274–80. https://doi.org/10.1021/acs.jnatprod.5b00709.

Lee, Hing-Ken, Kin-Fai Chan, Chi-Wai Hui, Ho-Kee Yim, Xun-Wei Wu, and Henry N. C. Wong. 2005. "Use of Furans in Synthesis of Bioactive Compounds." *Pure and Applied Chemistry* 77 (1):139–43. https://doi.org/10.1351/pac200577010139.

Leyva, Antonio, and Avelino Corma. 2009. "Isolable Gold(I) Complexes Having One Low-Coordinating Ligand as Catalysts for the Selective Hydration of Substituted Alkynes at Room Temperature without Acidic Promoters." *The Journal of Organic Chemistry* 74 (5):2067–74. https://doi.org/10.1021/jo802558e.

Lipshutz, Bruce H. 1986. "Five-Membered Heteroaromatic Rings as Intermediates in Organic Synthesis." *Chemical Reviews* 86 (5):795–819. https://doi.org/10.1021/cr00075a005.

Mascareñas, Jose L., and Fernando López. 2016. "Synthesis of Oxygenated Heterocyclic Compounds via Gold-Catalyzed Functionalization of π-

Systems." In *Topics in Heterocyclic Chemistry*, 10:1–52. https://doi.org/10.1007/7081_2015_5006.

Mukherjee, Rama, Vivek Kumar, Sanjay K Srivastava, Shiv K Agarwal, and Anand C Burman. 2006. "Betulinic Acid Derivatives as Anticancer Agents: Structure Activity Relationship." *Anti-Cancer Agents In Medicinal Chemistry* 6 (3):271–79. https://doi.org/10.2174/187152006776930846.

Nedopekina, Darya A., Rinat R. Gubaidullin, Victor N. Odinokov, Polina V. Maximchik, Boris Zhivotovsky, Yuriy P. Bel'skii, Veniamin A. Khazanov, Arina V. Manuylova, Vladimir Gogvadze, and Anna Yu. Spivak. 2017. "Mitochondria-Targeted Betulinic and Ursolic Acid Derivatives: Synthesis and Anticancer Activity." *Med. Chem. Commun.* 8 (10): 1934–45. https://doi.org/10.1039/C7MD00248C.

Omelchuk, O A, A S Tikhomirov, and A E Shchekotikhin. 2016. "Annelation of Furan Rings to Arenes." *Russian Chemical Reviews* 85 (8): 817–35. https://doi.org/10.1070/RCR4613.

Patil, Nitin T, and Yoshinori. Yamamoto. 2007. "Metal-Mediated Synthesis of Furans and Pyrroles." *Arkivoc* 2007 (10):121. https://doi.org/10.3998/ark.5550190.0008.a11.

Rudolph, Matthias, and A. Stephen K. Hashmi. 2012. "Gold Catalysis in Total Synthesis—an Update." *Chem. Soc. Rev.* 41 (6):2448–62. https://doi.org/10.1039/C1CS15279C.

Sarek, J, M Kvasnica, M Vlk, M Urban, P. Dzubak, and M. Hajduch. 2011. "The Potential of Triterpenoids in the Treatment of Melanoma." *Research on Melanoma - A Glimpse into Current Directions and Future Trends*, no. 1:125–58. https://doi.org/10.5772/19582.

Spivak, Anna Yu, Rinat R. Gubaidullin, Zulfiya R. Galimshina, Darya A. Nedopekina, and Victor N. Odinokov. 2016. "Effective Synthesis of Novel C(2)-Propargyl Derivatives of Betulinic and Ursolic Acids and Their Conjugation with β-d-Glucopyranoside Azides via Click Chemistry." *Tetrahedron* 72 (9):1249–56. https://doi.org/10.1016/j.tet.2016.01.024.

Spivak, Anna Yu, Jennifer Keiser, Mireille Vargas, Rinat R. Gubaidullin, Darya A. Nedopekina, Elvira R. Shakurova, Rezeda R. Khalitova, and

Victor N. Odinokov. 2014. "Synthesis and Activity of New Triphenylphosphonium Derivatives of Betulin and Betulinic Acid against Schistosoma Mansoni in Vitro and in Vivo." *Bioorganic & Medicinal Chemistry* 22 (21): 6297–6304. https://doi.org/10.1016/j.bmc.2014.07.014.

Spivak, Anna Yu, Darya A. Nedopekina, Rezeda R. Khalitova, Rinat R. Gubaidullin, Viktor N. Odinokov, Yuriy P. Bel'skii, Natalia V. Bel'skaya, and Veniamin A. Khazanov. 2017. "Triphenylphosphonium Cations of Betulinic Acid Derivatives: Synthesis and Antitumor Activity." *Medicinal Chemistry Research* 26 (3): https://doi.org/10.1007/s00044-016-1771-z.

Trofimov, B A, and E Yu Schmidt. 2014. "Reactions of Acetylenes in Superbasic Media. Recent Advances." *Russian Chemical Reviews* 83 (7):600–619. https://doi.org/10.1070/RC2014v083n07ABEH004425.

Urban, Milan, Jan Sarek, Miroslav Kvasnica, Iva Tislerova, and Marian Hajduch. 2007. "Triterpenoid Pyrazines and Benzopyrazines With Cytotoxic Activity." *Journal of Natural Products* 70 (4): 526–32. https://doi.org/10.1021/np060436d.

Vieser, Ralf;, and Wolfgang. Eberbach. 1995. "Studies on the Synthesis of Furans by Anionic Cyclization of 4-Pentynones." *Tetrahedron Letters* 36 (25):4405–8. https://doi.org/10.1016/0040-4039(95)00785-B.

Vitale, Paola, and Antonio Scilimati. 2013. "Five-Membered Ring Heterocycles by Reacting Enolates with Dipoles." *Current Organic Chemistry* 17 (18):1986–2000. https://doi.org/10.2174/13852728113179990093.

Wu, Jing, Bei-Hua Bao, Qi Shen, Yu-Chao Zhang, Qing Jiang, and Jian-Xin Li. 2016. "Novel Heterocyclic Ring-Fused Oleanolic Acid Derivatives as Osteoclast Inhibitors for Osteoporosis." *MedChemComm* 7 (2): 371–77. https://doi.org/10.1039/C5MD00482A.

Xu, Jun, Zhenxi Li, Jian Luo, Fan Yang, Ting Liu, Mingyao Liu, Wen-Wei Qiu, and Jie Tang. 2012. "Synthesis and Biological Evaluation of Heterocyclic Ring-Fused Betulinic Acid Derivatives as Novel Inhibitors of Osteoclast Differentiation and Bone Resorption." *Journal of Medicinal Chemistry* 55 (7): 3122–34. https://doi.org/10.1021/jm 201540h.

Zhdanko, Alexander, and Martin E. Maier. 2015. "Explanation of 'Silver Effects' in Gold(I)-Catalyzed Hydroalkoxylation of Alkynes." *ACS Catalysis* 5 (10): 5994–6004. https://doi.org/10.1021/acscatal.5b01493.

In: Furan: Chemistry, Synthesis and Safety ISBN: 978-1-53615-390-3
Editor: Ida Bailey © 2019 Nova Science Publishers, Inc.

Chapter 3

TITANIUM CATALYSIS IN THE CHEMISTRY OF FURANES

Leila O. Khafizova[*], *Mariya G. Shaibakova* *and Usein M. Dzhemilev*

Institute of Petrochemistry and Catalysis of Russian Academy of Sciences, Ufa, Russia

ABSTRACT

This chapter presents the research data published by the authors in the last five years on the new developed one-pot synthesis of tetrasubstituted furans by the reaction of symmetrical and unsymmetrical acetylenes with carboxylic acid esters and $EtAlCl_2$ in the presence of metallic magnesium (acceptor of chloride ions) and the Cp_2TiCl_2 catalyst.

The review considers the effect of the structure of initial acetylenes and carboxylic acid esters on the reaction direction and the yield of tetrasubstituted furans.

The reaction of symmetric acetylenes with esters of $α,ω$-dicarboxylic acids and $EtAlCl_2$ is described under the action of the Cp_2TiCl_2 catalyst. It was shown that the length of the hydrocarbon chain between the

---

[*] Corresponding Author's E-mail: Khafizovaleila@gmail.com.

carboxyl groups in α,ω-dicarboxylic acid esters has a specific effect on the chemoselectivity of the reaction. The Cp_2TiCl_2-catalyzed reaction of symmetrical acetylenes with α,ω-dicarboxylic acid esters in the presence of $EtAlCl_2$ can produce C_5-C_6 cyclic ketones containing alkylidene and alkenyl substituents.

Keywords: catalysis, Cp_2TiCl_2, symmetrical acetylenes, unsymmetrical acetylenes, $EtAlCl_2$, monocarboxylic acid esters, *α,ω*-dicarboxylic acid esters, tetrasubstituted furans

INTRODUCTION

Synthetic chemistry of heterocyclic compounds, in particular, the chemistry of furans represents a special section of organic chemistry, which is closely related to the problem of producing biologically active compounds.

The furan ring is an important structural block present in many natural products and drug molecules [1, 2]. Numerous synthetic furan derivatives were shown to exhibit various biological activities [3–7].

Furans have also been incorporated into polymers and macrocycles for materials applications [8, 9].

In addition, the utility of furans as building blocks in synthesis has received considerable attention [10–14].

Classical approach to furan synthesis is the Paal-Knorr method, in which 1,4-dicarbonyl compounds are transformed to furan derivatives [15–18].

Recently, several studies have focused on the development of metal-catalyzed transformation, including the cyclization of alkynyl [19–29], allenyl [30–38], cyclopropyl [39], and cyclopropenyl [40, 41] ketone derivatives. Alternative strategies include the cyclization of functionalized oxirane [42–47], alkynols [48–53], (Z)-2-en-4-yn-1-ols [54–58], substituted propargyl vinyl ethers [59–63], and others [64–72].

The elaboration of efficient methods for the synthesis of substituted furans using both widely known organic reactions and new methods based on available starting reagents and catalysts remains one of the important directions in the development of the chemistry of furans.

This chapter presents and discusses the results of research in the field of the synthesis of tetrasubstituted furans, based on the developed by the authors reaction of symmetrical and unsymmetrical acetylenes with EtAlCl$_2$, monocarboxylic and dicarboxylic acid esters catalyzed by Cp$_2$TiCl$_2$.

REACTION OF SYMMETRICAL ACETYLENES WITH ETHYL ALUMINUM DICHLORIDE AND ALKYL CARBOXYLIC ACID ESTERS

The idea of obtaining substituted furans was based on the assumption that acetylenes with EtAlCl$_2$ and carboxylic acid esters in the presence of metallic Mg and the Cp$_2$TiCl$_2$ catalyst can form intermediate 1,4-diones, which under reaction conditions will be then transformed into the corresponding furans in accordance with the Paal- Knorr reaction.

It should be noted that the idea described above was largely based on our previous findings while designing a method for the synthesis of alkoxy(hydroxy)cyclopropanes by the reaction of *alpha*-olefins with ethyl aluminum dichloride and carboxylic acid esters in the presence of the Cp$_2$ZrCl$_2$ catalyst [73, 74]. Bis(cyclopentadienyl)zirconacyclopropanes are the key intermediates in these reactions.

In accordance with this idea, we have assumed that the first stage of the reaction, under selected reaction conditions, in the presence of metallic Mg involves generation of metallocene "Cp$_2$M" (M = Zr, Ti) **1** (Scheme 1).

Complex **1** then undergoes the reaction with initial acetylene yielding, through the stage of formation of π-complex **2**, bis-cyclo penta dienyl zircona- or bis-cyclo pentadienyl titania cyclopropene **3** containing two active Zr-C (or Ti–C) bonds [75–78]. Then, intermediate **3** enters into reaction with appropriate carboxylic acid ester involving both active M−C bonds to afford dioxa metalla cycloheptane **4**, transmetallation of which *in* *situ* mediated by EtAlCl$_2$ can lead to oxadialuminums **5** with simultaneous regeneration of initial catalyst Cp$_2$MCl$_2$ (M = Zr, Ti). Migration of the alkoxide group in complex **5** to the aluminum atom followed by intramolecular rearrangement leads to 1,4-diketone **6**, which is transformed to the target tetrasubstituted furan **7**. Similar examples of furan formation have been reported [79–81].

Our preliminary experiments revealed that the reaction of oct-4-yne with EtAlCl$_2$ and ethyl acetate mediated by metallic Mg and Cp$_2$ZrCl$_2$ as the catalyst (oct-4-yne : [Al] : ester : Mg : [Zr] = 1 : 2 : 2 : 2 : 0.1, THF, 20°C, 48 h), after hydrolysis of the reaction mixture, provides the formation of 2,5-dimethyl-3,4-dipropylfuran **8** (40% yield) together with two by-products, namely, 1,2,3,4-tetrapropylbutadiene **9** and hexapropylbenzene **10** as the acetylene trimerization product (3:3:1 ratio) in 95% total yield (Scheme 2). As known, product **9** can be derived from 1-ethyl-2,3,4,5-tetrapropylaluminacyclopentadiene resulting from the catalytic cycloalumination of the starting oct-4-yne [82].

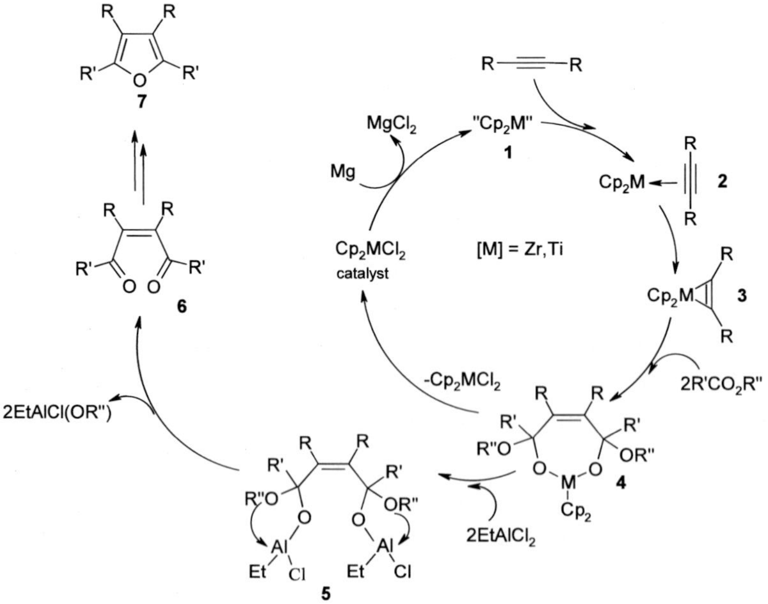

Scheme 1.

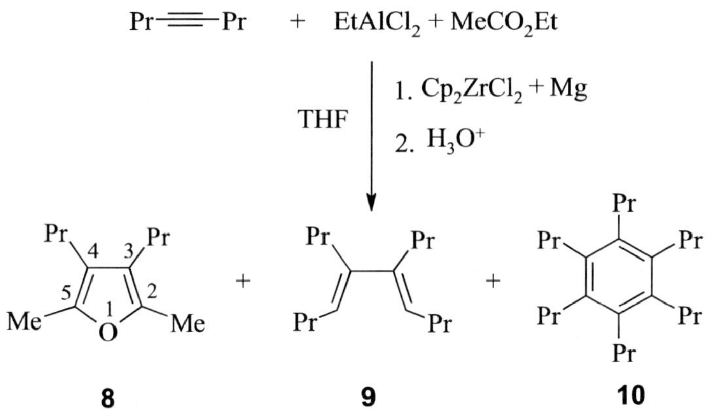

Titanium Catalysis in the Chemistry of Furanes

Pr≡Pr + EtAlCl$_2$ + MeCO$_2$Et

THF | 1. Cp$_2$ZrCl$_2$ + Mg
    | 2. H$_3$O$^+$

↓

8 (2,5-dimethyl-3,4-dipropylfuran) + **9** (diene) + **10** (hexapropylbenzene)

95% yield (**8** : **9** : **10** = 3 : 3 : 1), 20 °C, 48 h

Scheme 2.

To improve the yield of tetraalkyl substituted furan **8** we have optimized the reaction parameters by varying the catalyst and solvent nature, temperature and duration of reaction as well as the ratio of reactants.

It was found that the highest yield of 2,5-dimethyl-3,4-dipropylfuran **8** (48%) can be achieved at a molar ratio of oct-4-yne : [Al] : ethyl acetate: Mg: [Zr] = 1 : 2 : 2 : 2 : 0.1 and by carrying out the reaction at 60°C for 6 hours in tetrahydrofuran.

The elevated temperature (60°C) greatly reduces the reaction time from 48 h to 6 hours.

Among the catalysts tested in the reaction of oct-4-yne with EtAlCl$_2$ and ethyl acetate (Table 1.1), the Cp$_2$TiCl$_2$ catalyst exhibited the highest catalytic activity.

The most surprising effect was observed, when Cp$_2$ZrCl$_2$ has been replaced by 10 mol% Cp$_2$TiCl$_2$. With this catalyst, the highest yield of 2,5-dimethyl-3,4-dipropylfuran **8** were achieved 80% under optimal reaction conditions (THF, 60°C, 6 h, oct-4-yne : [Al] : ester : Mg : [Ti] = 1 : 2 : 2 : 2 : 0.1). The combined yield of byproducts **9** and **10** under these reaction conditions did not exceed 15%. Without catalyst the reaction between oct-4-yne and ethyl acetate mediated by EtAlCl$_2$ does not occur.

Table 1.1. Effect of the catalyst nature on the yield of furan 8 in the reaction of oct-4-yne with ethyl acetate mediated by EtAlCl$_2$

Entry	Catalyst	Yield of 8, %
1	Cp$_2$TiCl$_2$	80
2	Cp$_2$ZrCl$_2$	48
3	VO(acac)$_2$	25
4	Zr(acac)$_4$	32
5	Ti(acac)$_2$Cl$_2$	44
6	Cp$_2$HfCl$_2$	0
7	Ti(OPr^i)$_4$	12
8	TaCl$_5$	8
9	Without catalyst	-

We have found that the solvent plays a significant role in this reaction, which occurs only in tetrahydrofuran. The use of another solvent such as hexane, benzene, toluene, the diethyl ether, and methylene chloride suppressed the furan formation (Table 1.2).

The increase in concentration of EtAlCl$_2$ or the ester towards the starting acetylene in the reaction mixture does not favor the enlargement of the total yield of the target tetrasubstituted furans.

Table 1.2. Effect of the solvent nature on the yield of furan 8 in the reaction of oct-4-yne with ethyl acetate mediated by EtAlCl$_2$ and Cp$_2$TiCl$_2$

Entry	Solvent	Yield of 8, %
1	THF	80
2	Diethyl ether	25
3	1,4-Dioxane	10
4	Methylene dichloride	2
5	Hexane	-
6	Cyclohexane	-
7	Toluene	-
8	Benzene	-

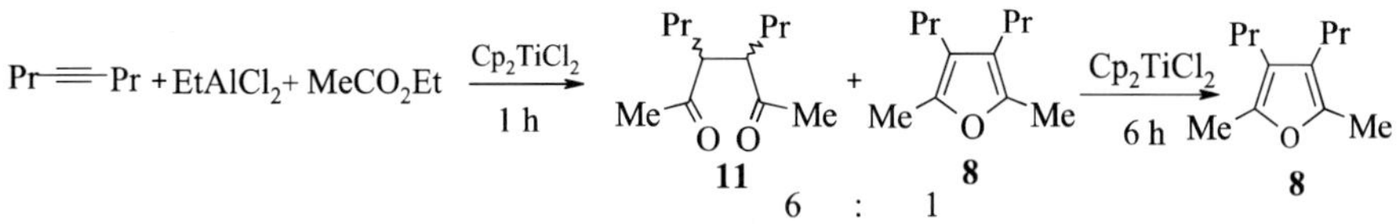

Scheme 3.

Concentration of the catalyst significantly affects the yield of 2,5-dimethyl-3,4-dipropylfuran **8**. Thus, the highest yield of **8** (80%) was achieved at the concentration of Cp$_2$TiCl$_2$ equal to 10 mol% at 60°C. With a decrease in the concentration of the catalyst to 1 mol% the yield of tetrasubstituted furan **8** decreased to 20%. This is probably due to a decrease in the concentration of catalytically active sites in the reaction mixture.

The structure of 2,5-dimethyl-3,4-dipropylfuran **8** was proved by means of modern spectral methods such as one-dimensional (^{1}H, ^{13}C) and two-dimensional (HSQC, HMBS) NMR, IR, UV and also using mass spectrometry.

The ^{13}C NMR spectrum of furan **8** contains two resonances at δC 144.5 and 119.2 belonging to the C(2,5) and C(3,4) atoms of the furan ring respectively. Additionally, the spectrum exhibits the resonance at δC 11.63 attributable to the alpha-methyl groups attached to the furan C(2) and C(5) atoms. In the HSQC experiment this signal correlates to the singlet signal at δH 2.17. The position of the methyl groups were determined in accordance with observations in the HMBS spectra.

Our experiments have shown that the reaction between oct-4-yne, ethyl acetate and EtAlCl$_2$ produces 3,4-dipropylhexane-2,5-dione **11** together with tetrasubstituted furan **8** (~ 6:1 ratio) under selected reaction conditions after 1 h reaction time. With increasing the reaction time from one to six hours, the concentration of dione **11** in the mixture decreased, while the concentration of furan **8** increased. After 6 h, dione **11** completely transformed to the desired furan **8** in 80% yield (Scheme 3).

The ^{13}C and ^{1}H NMR spectra of 3,4-dipropylhexane-2,5-dione **11** contain two sets of signals indicating the presence of *d,l* and *meso* diastereomers in the ratio 2:3. The assignment of signals for each of stereomers was made on the basis of spectral criteria described for 3,4-dialkyl-2,5-diones [83]. Chemical shifts of the carbon atoms in *meso* isomer located in positions 2 and 3 and also in α-position of the alkyl substituent appear in a lower field of the NMR spectra relative to the corresponding signals for isomer with *d,l*-configuration of asymmetric centers. Significant difference of the proton chemical shifts of the methylene group located in α-position to the asymmetric centers for *meso* diastereomer is the additional evidence for executed stereochemical identification. Magnetic equivalence of the appropriate methylene protons in *d,l*-diastereomer is due to the existence of symmetrical forms at free rotation around the C^2-C^3 bond.

We have performed the reaction of EtAlCl$_2$ with **11** isolated from the reaction mixture through one hour after the start of the reaction between oct-4-yne, EtAlCl$_2$ and ethyl acetate under the same reaction conditions (Cp$_2$TiCl$_2$, Mg, THF, 60°C, 6 h). As a result, 2,5-dimethyl-3,4-dipropylfuran **8** has been obtained in 80% yield thus reliably supporting the participation of 1,4-diones as the key intermediates upon formation of tetrasubstituted furans.

Later we have found that the developed reaction provides formation of furans **12a** and **12b** as it takes place in the case of other esters of alkyl carboxylic acids, for example, ethyl propionate, propyl propionate and isoamyl butyrate (Scheme 4).

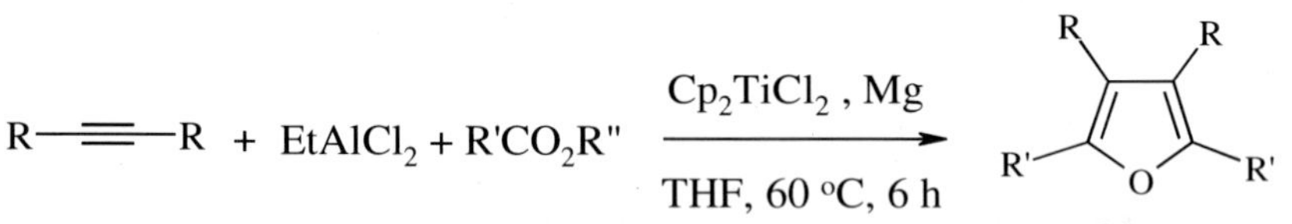

$$R\!\!\equiv\!\!R + EtAlCl_2 + R'CO_2R'' \xrightarrow[\text{THF, 60 °C, 6 h}]{Cp_2TiCl_2,\ Mg}$$

12a–e

a R = Pr, R' = Et, R" = Et;
b R = Pr, R' = Pr, R" = Am^i;
c R = Et, R' = Me, R" = Et;
d R = Et, R' = Et, R" = Et;
e R = Bu, R' = Me, R" = Et

Scheme 4.

In an analogous fashion, symmetrical acetylenes, such as hex-3-yne and dec-5-yne, react with EtAlCl$_2$ and alkyl carboxylic acid esters in the presence of the Cp$_2$TiCl$_2$ catalyst giving rise to appropriate tetraalkyl substituted furans **12c**, **12d** and **12e** under similar reaction conditions with quantitative yields.

REACTION OF SYMMETRICAL ACETYLENES WITH ETHYL ALUMINUM DICHLORIDE AND CYCLOALKYL(ARYL, HETARYL)CARBOXYLIC ACID ESTERS

In order to extend the scope of the reaction between symmetrical acetylenes, EtAlCl$_2$ and the simplest esters of aliphatic carboxylic acids, as well as elucidate the effect of the initial ester structure on the reaction pathway, we examined analogous reactions with cyclohexanecarboxylic **13**, cyclopentanecarboxylic **15**, cyclobutanecarboxylic **16**, thiophene-2-carboxylic **17**, thiophen-2-ylacetic **18**, benzoic **21**, phenylacetic **22**, cyclohexylacetic **14**, furan-2-carboxylic **19**, and furan-2-ylacetic **20** acid esters.

It was established that the reactions of symmetrical internal acetylenes, oct-4-yne and dec-5-yne, with cycloalkylcarboxylic acid esters **13–16** and EtAlCl$_2$ in the presence of the Cp$_2$TiCl$_2$ catalyst under developed reaction conditions (acetylene : [Al] : ester : Mg : [Ti] = 1 : 2 : 2 : 2 : 0.1, THF, 6 h, 60°C [84] afforded tetrasubstituted furans **23–26** in 72 – 82% yield (Scheme 5).

The investigation into the influence of aromatic and heteroaromatic substituents in the structure of carboxylic acid esters on the reaction of the furan formation has shown that thiophene-2-carboxylic **17**, furan-2-carboxylic **19**, and benzoic **21** acid esters, in which the carbonyl group is conjugated with the aromatic or heteroaromatic ring, failed to react with symmetrical acetylenes under the given conditions.

By contrast, thiophen-2-ylacetic **18**, furan-2-ylacetic **20**, and phenylacetic **22** acid esters, where the ester group is separated from the aromatic ring by the methylene unit reacted with EtAlCl$_2$ and symmetrical acetylenes under developed reaction conditions to give 38 – 52% of the corresponding tetrasubstituted furans **27**, **28**, and **29** (Scheme 6).

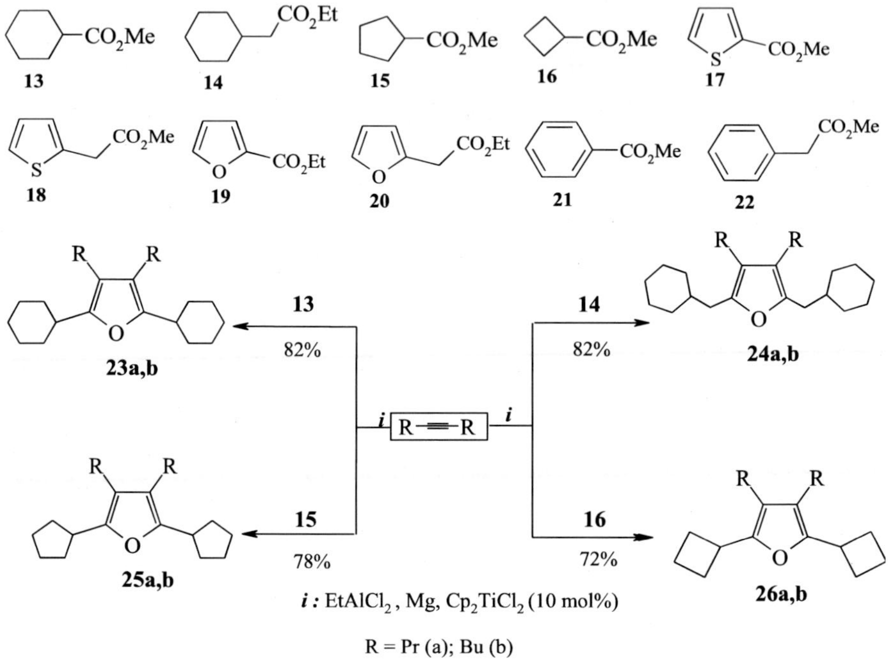

Scheme 5.

The low reactivity of aromatic and heteroaromatic carboxylic acid α-esters **17**, **19**, and **21** unlike β-esters **18**, **20**, and **22** as well as cycloalkanecarboxylic acid esters **13**, **15**, and **16** is likely to be determined by mesomeric effect of the aromatic (or heteroaromatic) ring relative to the ester carbonyl group.

Upon introduction of the methylene group between the aromatic ring and ester group mesomeric effect weakens in going to aryl(hetaryl)acetic acid esters, and therefore thiophen-2-ylacetic **18**, furan-2-ylacetic **20**, and phenylacetic **22** acid esters are readily converted into furan derivatives, though in lower yields than in the reactions with cycloalkanecarboxylates.

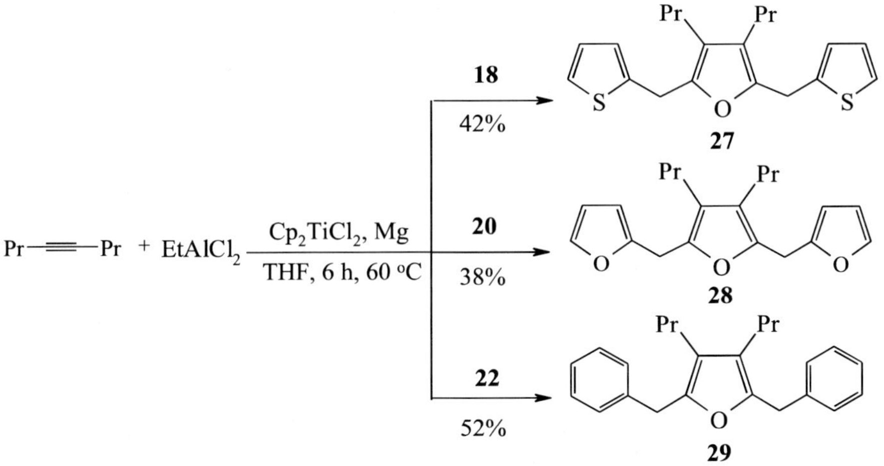

Scheme 6.

To test the proposed assumption about the influence of the conjugation effect on the reactivity of the carbonyl group in the structure of aryl and heteroaryl carboxylic acid esters we have studied the reaction of oct-4-yne with ethyl acrylate and ethyl hept-6-enoate in the presence of EtAlCl$_2$ and Cp$_2$TiCl$_2$ as the catalyst.

The obtained results evidenced that no furan derivative was obtained from ethyl acrylate, whereas the reaction with ethyl hept-6-enoate afforded substituted furan 30 in 80% yield (Scheme 7).

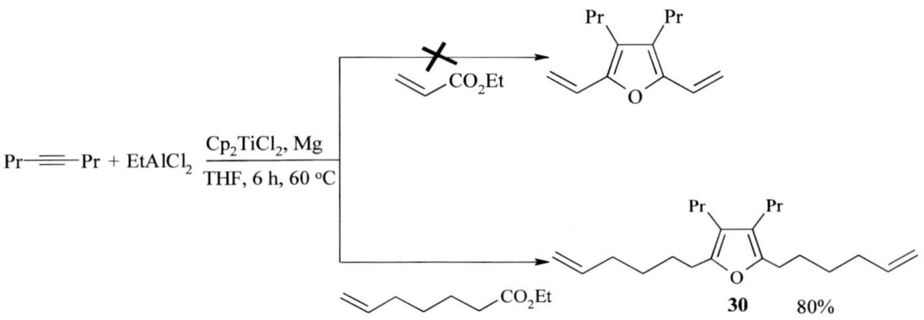

Scheme 7.

The structure of the obtained furans **23–30** was proved using modern physicochemical methods of analysis [one-dimensional (^{1}H, ^{13}C, Dept 135) and two-dimensional (HSQC, HMBC, and H–H COSY) NMR techniques, IR spectroscopy and gas chromatography mass spectrometry (GC/MS)].

Thus, C_4–C_6-cycloalkanecarboxylic acid esters readily react with symmetrical acetylenes and $EtAlCl_2$ in the presence of Cp_2TiCl_2 as the catalyst to produce tetrasubstituted furans. Aromatic and heteroaromatic carboxylic acid esters are inactive in this reaction. Aryl- and hetarylacetic acid esters, in which the carbonyl group is not conjugated with the aromatic ring, do react under the developed reaction conditions, yielding tetrasubstituted furans. An advantage of the proposed procedure is the possibility for synthesizing difficultly accessible but practically important furan derivatives in one preparative step from commercially available symmetrical acetylenes and carboxylic acid esters.

REACTION OF SYMMETRICAL ACETYLENES WITH ETHYL ALUMINUM DICHLORIDE AND CYCLOPROPANECARBOXYLIC ACID ESTER

Methyl ester of cyclopropanecarboxylic acid among the studied esters of cycloalkane (cyclobutane-, cyclopentane-, cyclohexane-)carboxylic acids occupies a special place. Our further studies have shown that, in contrast to C_4–C_6 cycloalkanecarboxylates, methylcyclopro-pane-carboxylate under the reaction conditions (molar ratio acetylene:[Al]: ester : Mg : [Ti] =1 : 2 : 2 : 2 : 0.1, THF, 6 h, 60°C) gives rise to 2,3-dialkyl-1,4-dicyclopropyl-1,4-diketones **31a**, **31b**, and **31c** in 62–66% yield (Scheme 8).

We have found that 1,4-diketones **32** were the close precursors of furans **33** in the reaction of symmetrical acetylenes with $EtAlCl_2$ and carboxylates catalyzed by Cp_2TiCl_2 (Scheme 9) [85]. This is in accordance with the mechanism of Paal–Knorr reaction in the presence of Lewis acids [17, 18]. Thus, for example, in the sample taken in 1 h after the beginning of the reaction, 1,4-diketones **32a,b** (**33a,b** : **32a,b** = 3 : 2) were detected, along with substituted furans **33a,b** and starting materials. Compounds **32a** and **32b** were completely transformed into furans **33a** and **33b** within 6 h.

In order to examine the dependence of chemoselectivity of the cyclization reaction of 1,4-diketones into the corresponding furans on the nature of the starting substrate, we have investigated potential energy surface of the reactions involving two diketones (1,4-dicyclopropyl-2,3-dipropylbutane-1,4-dione and 1,4-dicyclobutyl-2,3-dipropylbutane-1,4-dione) and $EtAlCl_2$ using quantum chemical methods. The detailed theoretical study of the mechanism for the formation of the five-membered heterocycles from 1,4-dicarbonyl compounds (the *Paal–Knorr reaction*) has been earlier undertaken by B. Mothana and R. J. Boyd on the example of the pyrrole synthesis [86]. The proposed mechanism, based on the kinetics studies of the reactions of 3,4-dimethylhexane-2,5-dione with the primary amines [87, 88], and acids [89], leading to substituted pyrroles and furans, was consistent with the experimental data.

To clarify the observed chemoselectivity we have studied the key stage of the cyclization reaction of 1,4-dicyclobutyl-2,3-dipropylbutane-1,4-dione **32b**.

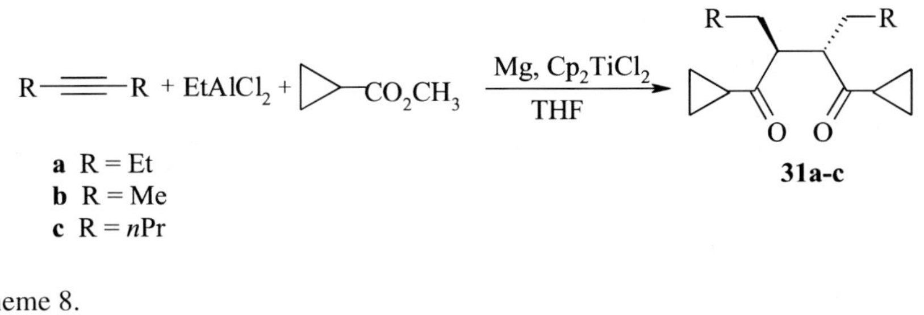

Scheme 8.

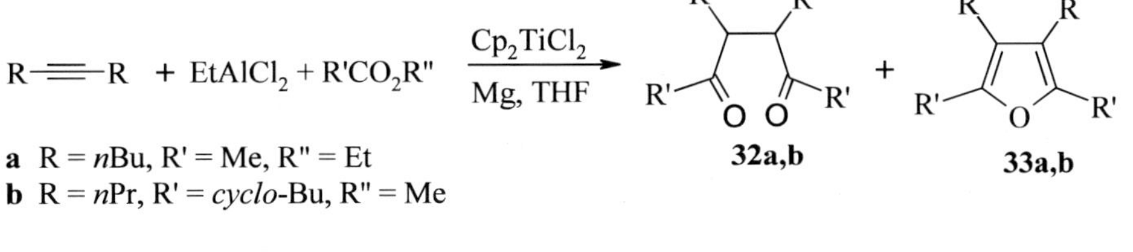

Scheme 9.

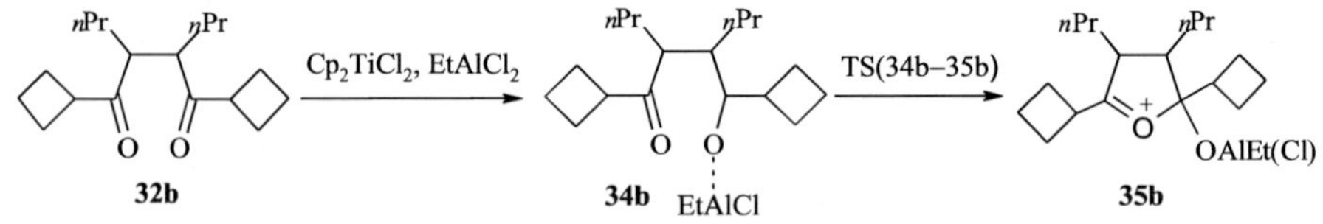

Scheme 10.

Considering the literature data [84, 86, 90, 91], we hypothesized that the interaction between **32b** and Cp$_2$TiCl$_2$ catalyst followed by subsequent transmetallation of the resulting adduct with EtAlCl$_2$ leads to the formation of intermediate **34b** (Scheme 10).

DFT studies have shown that the transformation of **34b** to **35b** represents the thermodynamically favorable process since the calculated Gibbs energy for this reaction is negative (ΔG = − 26.4 kcal/mol). Scanning the potential energy surface along the cyclization reaction of **32b** revealed the presence of maximum, whose corresponding optimized structure is represented in Figure 1. The activation barrier for this reaction was low ($\Delta G \neq$ = 11.4 kcal/mol). These observations confirm the possibility of forming the furan cycle **35b**.

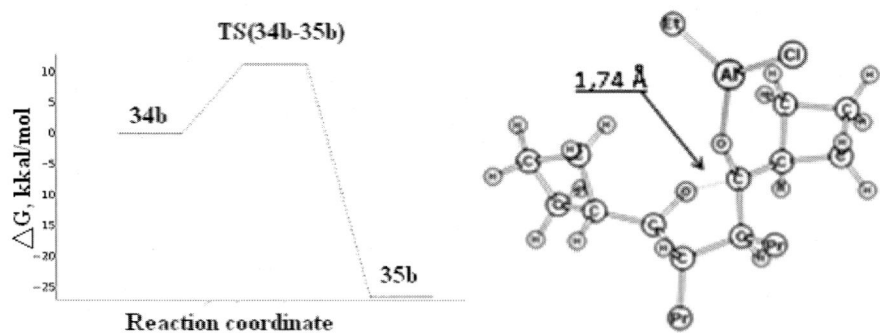

Figure 1. The energy diagram of the reaction 34b→35b and DFT/PBE optimized structure for the TS (34b–35b) (ν = 362.8i cm^{-1}).

However, it proved impossible to localize such a transition state and minimum of type **35b** for the process involving 1,4-dicyclopropyl analogue **31a**. Apparently, cyclopropyl substituents essentially accelerate solvolytic processes thus preventing further transformations of **31a** to form **35a** [92, 93]. Ring opening decelerates solvolysis and thus makes formation of furan possible. The numerous reactions, where the cyclopropane ring opening is preceded by prior reaction with Lewis acids, are well-known and proceed smoothly [94–100].

We have carried out the reaction between **31a** and $AlCl_3$ under strong reaction conditions in the presence of a mixture of concentrated hydrochloric and acetic acids [80]. As a result, 2,5-bis(3-chloropropyl)-3,4-dipropylfuran **36** was obtained in 92% yield (Scheme 11).

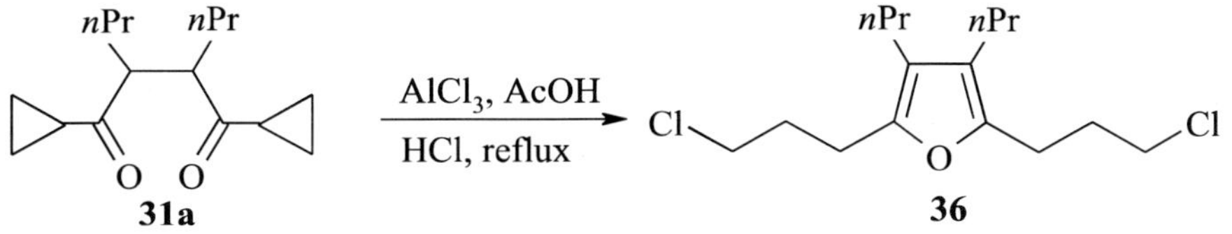

Scheme 11.

Comparative quantum chemical studies of the cyclization stage of aluminum-containing intermediates involving 1,4-dicyclopropyl-2,3-dipropyl-butane-1,4-dione **31a** and 1,4-dicyclobutyl-2,3-dipropyl-butane-1,4-dione **32b** have shown that the reaction cannot bring about adducts bearing cyclopropyl substituents that allows one to explain chemoselectivity of the reaction studied.

REACTION OF SYMMETRICAL ACETYLENES WITH ETHYL ALUMINUM DICHLORIDE AND ESTERS OF DICARBOXYLIC ACIDS

In continuation of our investigations aimed at clarifying the possibility to conduct the reactions utilizing esters of $α,ω$-dicarboxylic acids, we have studied the reaction between symmetrical acetylenes and various aliphatic dicarboxylic acid diesters of oxalic, malonic, succinic, glutaric, adipic, pimelinic, cork and azelaic acids under the reaction conditions previously developed [101] (acetylene : [Al] : ester : Mg : [Ti] =1 : 2 : 2 : 2 : 0.1, THF, 6 h, 60°C) for monocarboxylic acids.

The first homologues of the series of esters, namely, methyl esters of oxalic, malonic, and succinic acids were unreactive with symmetrical acetylenes and EtAlCl$_2$ in the presence of Cp$_2$TiCl$_2$. All attempts to implement these reactions by varying the reaction conditions (e.g., reaction time, catalyst loading, temperature, reagent ratios) were unsuccessful. Apparently, the above dicarboxylates act as bidentate ligands to form stable complexes with EtAlCl$_2$ and the central Ti atom in the catalyst, thus blocking the formation of the desired furans.

Unexpected results were obtained with methyl esters of adipic and glutaric acids. Thus, the reaction between symmetrical acetylenes (oct-4-yne or dec-5-yne) and the methyl ester of glutaric acid led to the corresponding 2,3-dialkylcyclohept-2-en-1-ones **37a,b** and alkenylcyclopentan-1-ones **38a,b** (2:1 molar ratio) in 68 − 70% total yield (Scheme 12). We have also isolated very small amounts (less than 10%) of 2,3-dialkylcyclohept-2-ene-1,4-diones **39a** and **39b** from the reaction mixture.

In an analogous fashion, dimethyl adipate gave 2-alkylidene cyclohexan-1-ones **40a,b** and 2-alkenyl cyclohexan-1-ones **41a,b** (40 : 41 = 2 : 1 molar ratio) in 63−68% total yield (Scheme 13).

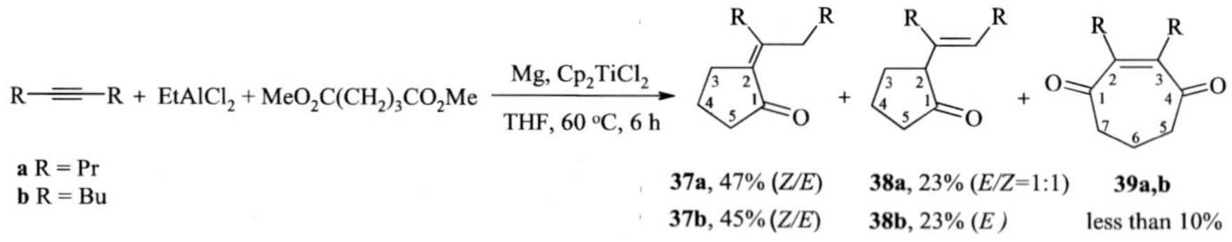

R≡R + EtAlCl$_2$ + MeO$_2$C(CH$_2$)$_3$CO$_2$Me $\xrightarrow{\text{Mg, Cp}_2\text{TiCl}_2}$ THF, 60 °C, 6 h

a R = Pr
b R = Bu

37a, 47% (Z/E)   38a, 23% (E/Z=1:1)   **39a,b**
37b, 45% (Z/E)   38b, 23% (E)   less than 10%

Scheme 12.

R≡R + EtAlCl$_2$ + MeO$_2$C(CH$_2$)$_4$CO$_2$Me $\xrightarrow{\text{Mg, Cp}_2\text{TiCl}_2}$ THF, 60 °C, 6 h

a R = Pr
b R = Bu

40a, 43% (Z/E)   41a, 25% (E/Z = 4:1)
40b, 42% (Z/E)   41b, 21% (E/Z = 10:1)

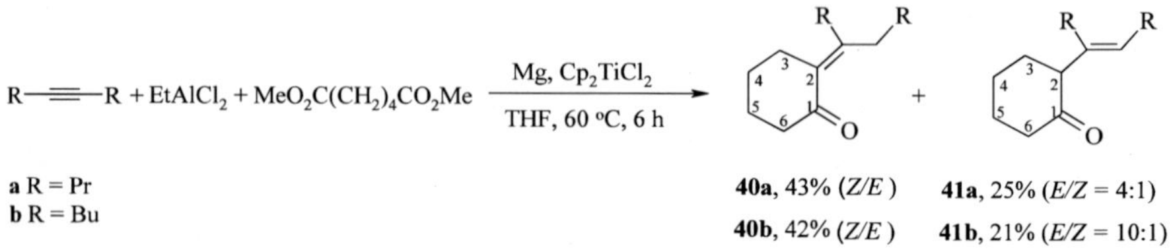

Scheme 13.

We have found that the reaction of symmetrical acetylenes with dimethyl glutarate and dimethyl adipate under developed reaction conditions provided E/Z mixtures **37a,b** и **40a,b** (~ 1:1 according to the ^{13}C NMR data). Analysis of signal intensities of the characteristic protons on the double bond in the ^{1}H NMR spectra indicated that *E* isomer is predominant in the case of **38b** and **41a,b** (*E/Z* = 4:1 for **41a**, and *E/Z* = 10:1 for **41b**), while *E/Z* = 1:1 for **38a**.

The stereochemistry of the double bond was established from NOESY experiments on the example of *Z/E* stereoisomers **38a**, as well as the isolated individual *E* isomer **38b**. Thus, the ^{1}H NMR spectrum of the isomeric mixture of **38a** reveals two triplets (δH 5.2 and 5.4 ppm) corresponding to the methylene protons for two isomers. The NOESY spectrum demonstrates a cross-peak between the closely spaced protons, namely, between the signal of protons on the double bond at δH 5.2 ppm and the tertiary proton in the ring (δH 2.7 ppm). For the second isomer no similar interactions were found. Our experiments have shown that for the series of compounds **41a,b** there also exists a NOE-effect between the proton on the double bond at 5.1 ppm and the proton on the asymmetric carbon C_2. This allowed to classify compounds **41a,b** as geometric isomers, namely, *E* isomers.

All the data from the experiments are in a very good agreement with the theory. Thus, the potential energy surface scan (PES scan) using DFT/B3LYP method by rotating the C_2–CH= group allowed to locate three minima, which correspond to stable rotamers for each isomer. We have established that only for *E* isomer in *ap*-conformation the distance between the said protons does not exceed 3.0 Å (Figure 2) thus favoring NOE interactions.

It should be noted that the hexanone ring in the solution predominantly exists in a chair conformation with the substituent in equatorial position as evidenced by the values of vicinal H-H coupling constants ($^3J_{aa}$ = 12.0 and $^3J_{ae}$ = 5.3 Hz) for the methine proton. Thus, based on a complex analysis of the theoretical and experimental data obtained, reliable assignments of isomers were made.

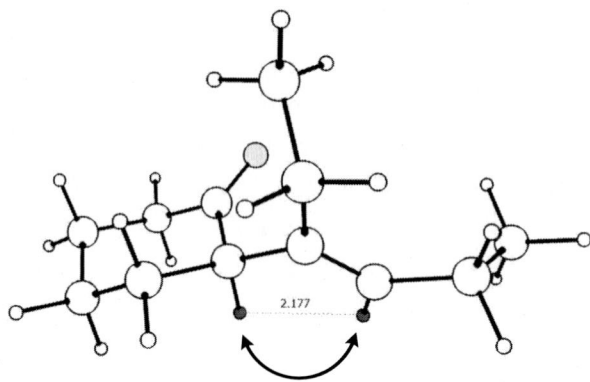

Figure 2. The optimized structure of the stable ap-rotamer of the E isomer with the closely spaced protons, for which the NOE interaction is possible, on the example of model 2-[1-ethylbut-1-en-1-yl]-cyclohexanone.

Concerning the mechanism for the formation of C_5-C_6 cyclic ketones, we presume that the above compounds are produced through titanacyclopropene intermediates, which react with the corresponding dicarboxylates involving Ti-C bonds. Further transformations giving alkylidene and alkenyl substituted ketone products remain unclear. The mechanism for the formation of products of the reaction of symmetric acetylenes with dicarboxylic acid ester and ethylaluminum dichloride in the presence of metallic magnesium and Cp_2TiCl_2 catalyst seems to us a rather complex catalytic process, which involves the simultaneous occurrence of a series of parallel reactions. Additional investigations are being carried out in order to get new experimental results that can clarify these aspects.

The least complicated example, in our opinion, is the reaction leading to 1,4-diones with the participation of diester of glutaric acid. The proposed scheme is presented below (Scheme 14).

The key stage of this reaction is the formation of titanacyclopropene intermediate **I**. The insertion of glutaric acid diester occurs into one Ti-C bond of the intermediate **I** to form the corresponding oxotitanacyclopropene **II**. Migration of two alkoxy groups to the titanium atom and subsequent elimination of the $Cp_2Ti(OMe)_2$ molecule results in a synchronous ring closure to form 1,4-diones **39**.

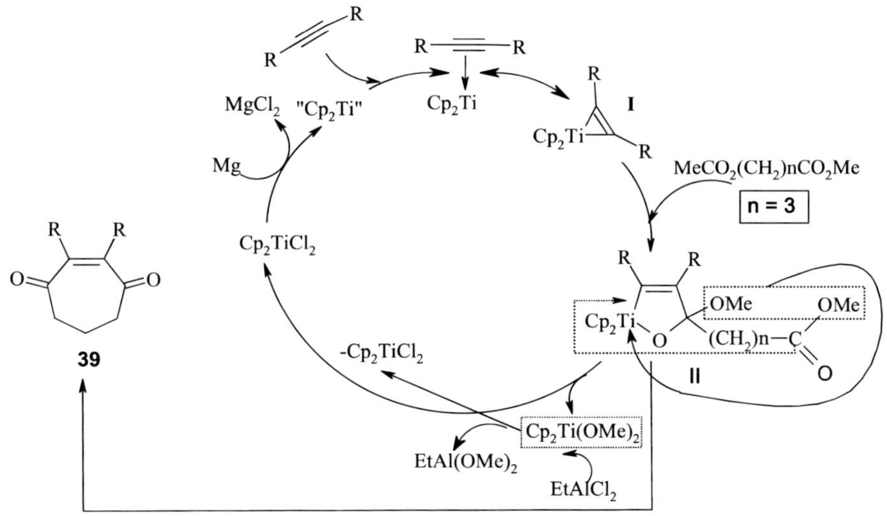

Scheme 14.

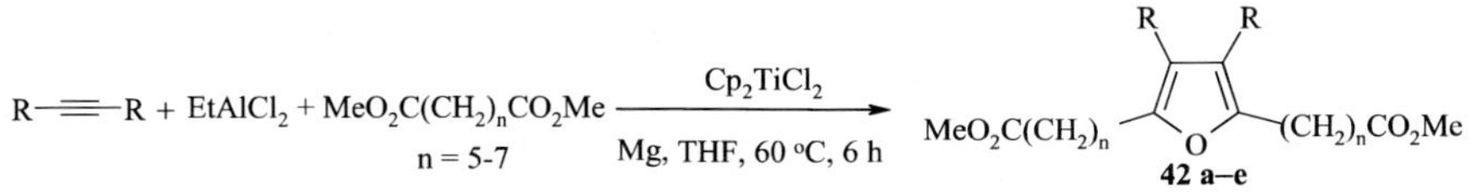

Scheme 15.

a R = Pr, n = 5 (65%); **b** R = Et, n = 6 (65%); **c** R = Pr, n = 6 (65%); **d** R = Bu, n = 6 (62%); **e** R = Et, n = 7 (60%)

We have examined the behavior of the long-chain carboxylic esters in the reaction under study. As in the case of monocarboxylic acid esters, the reactions with diesters of pimelic, cork and azelaic acids afford tetrasubstituted furans **42a-e** in 60-65% yield (Scheme 15).

These observations suggest that the length of the hydrocarbon linker between the dicarboxylate carbonyl groups significantly affects the reaction pathway. Chemoselectivity of the Cp_2TiCl_2-catalyzed reaction of symmetrical acetylenes with $EtAlCl_2$ and diesters of dicarboxylic acids depends upon the length of the hydrocarbon chain between the carboxyl groups in α,ω-dicarboxylate.

The investigations have shown that only diesters of α,ω-dicarboxylic acids, in which the hydrocarbon chain between the carbonyl groups exceeds two CH_2 groups, enter the reaction with acetylenes in the presence of $EtAlCl_2$ and the Cp_2TiCl_2 catalyst. Esters of $α,ω$-dicarboxylic acids (e.g., pimelic, cork, and azelaic), in which the hydrocarbon chain between the carboxyl groups exceeds four CH_2 groups, form tetrasubstituted furans. In the case of glutaric and adipic acid diesters, in which the hydrocarbon chain between the carboxyl groups equal to three or four CH_2 groups, form alkylidene and alkenyl substituted C_5–C_6 cyclic ketones.

UNSYMMETRICAL ACETYLENES IN TI-CATALYZED REACTIONS WITH ETHYL ALUMINUM DICHLORIDE AND ESTERS OF MONOCARBOXYLIC ACIDS

In order to develop one-pot methods for the synthesis of promising substituted furans, we have investigated the reaction between unsymmetrical acetylenes including those with functional groups and esters of monocarboxylic acids in the presence of $EtAlCl_2$ and Cp_2TiCl_2 as the most efficient catalyst in this reaction.

We have previously shown [101, 102] that upon interaction between symmetric acetylenes, $EtAlCl_2$ and esters of aliphatic carboxylic acids in the presence of the Cp_2TiCl_2 catalyst 1,4-butanediones are the key intermediates, which under selected conditions according to Paal-Knorr reaction in the presence of Lewis acids undergo rearrangement to corresponding tetrasubstituted furans.

The experiments with dialkyl substituted unsymmetrical acetylenes demonstrated that under the reaction conditions (acetylene : [Al] : ester : Mg : [Ti] = 1 : 2 : 2 : 2 : 0.1, THF, 6 h, 60°C), as in the case of symmetrical acetylenes, the formation of 1,4-diketone intermediates also occurs, which are easily *in situ* transformed into tetrasubstituted furans. Thus, the reaction between hept-2-yne (or oct-2-yne), ester of acetic (propionic, butyric) acid and $EtAlCl_2$, produces 1,4-diones **43a–d** together with tetrasubstituted furans **44a–d** (~ 6:1 ratio) under selected reaction conditions after 1 hour reaction time. With increasing the reaction time from one to six hours, the concentration of **43a–d** in the mixture decreased, while the concentration of **44a–d** increased. After 6 hours, the desired furans **44a–d** were obtained in 70 - 75% yield (Scheme 16).

Further experiments with unsymmetrical acetylenes (oct-2-yne, methylcyclopropylacetylene) revealed that the replacement of aliphatic carboxylic acid esters by methyl ester of cyclopropanecarboxylic acid under selected reaction conditions does not result in the formation of desired furans as in the case of symmetrical acetylenes. Instead, exclusively 2,3-disubstituted 1,4-dicyclopropyl-1,4-diones **45** and **46** were obtained in 62 – 66% yield (Scheme 17).

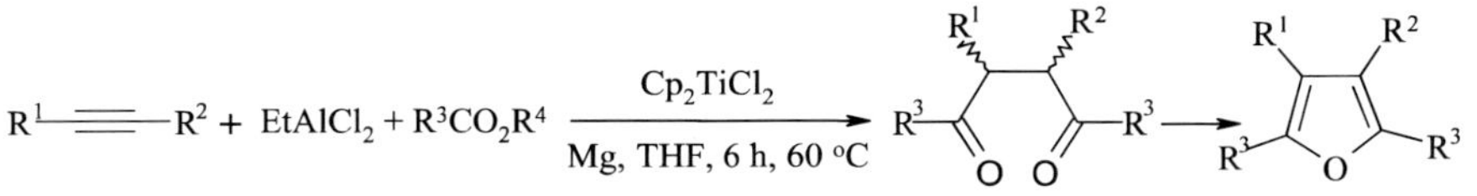

1a R^1 = Me, R^2 = nAmyl, R^3 = R^4 = Et; **1b** R^1 = Me, R^2 = nAmyl, R^3 = R^4 = Me;
1c R^1 = Me, R^2 = nAmyl, R^3 = Pr, R^4 = Me; **1d** R^1 = Me, R^2 = Bu, R^3 = R^4 = Et

Scheme 16.

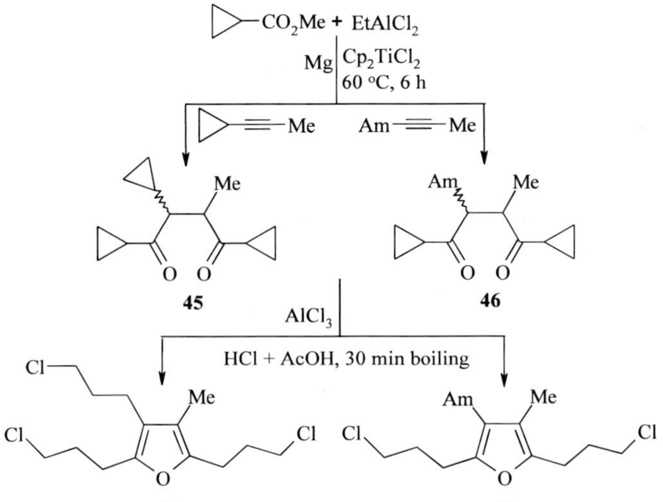

Scheme 17.

Considering the results of our studies we have hypothesized that furans can be formed from 1,4-dicyclopropyl-1,4-diones **45** and **46** after cyclopropane ring opening. Numerous examples are known in the literature, where protic and aprotic acids are used for cleavage of the cyclopropane ring [94–100].

To confirm our assumptions and cyclize diketones **45** and **46** to furans (Paal-Knorr reaction), we have performed the above reaction followed by adding a mixture of $AlCl_3$, concentrated hydrochloric and acetic acids (Scheme 17) [80]. Under these reaction conditions diones **45** and **46** completely transformed to 4-methyl-2,3,5-tris(3-chloropropyl)-furan **47** and 3-methyl-4-pentyl-2,5-bis(3-chloropropyl)-furan **48** thus revealing low reactivity of 1,4-dicyclopropyl-1,4-diones in the Paal-Knorr reaction (Scheme 17).

To expand the scope of application of the developed reaction and for analyzing the effect of substituents in unsymmetrical acetylenes on the formation of tetrasubstituted furans, a number of other acetylenes, namely, methylphenylacetylene, methylcyclopropylacetylene, 4-oct-2-yn-1-yl-3,6-dihydro-2H-pyran, and 6-chloro-hex-2-yne, have been examined in the reaction with ethyl acetate and $EtAlCl_2$ catalyzed by Cp_2TiCl_2. We have found that above unsymmetrical acetylenes produce predominantly 2,3-disubstituted 1,4-diones **50**, **52**, **54**, and **56** together with tetrasubstituted furans **49**, **51**, **53**, and **55** (~ 2:1 ratio) in 65-85% total yield (Scheme 18). A further increase in the reaction time and temperature did not alter the yield and ratio of products.

It became clear that unsymmetrical acetylenes having aromatic, cycloalkyl, heterocycloalkyl, and haloalkyl substituents enter into the multicomponent reaction with esters of alkyl carboxylic acids and $EtAlCl_2$ in the presence of the Cp_2TiCl_2 catalyst and metallic magnesium to afford a product mixture, which contains tetrasubstituted furans and corresponding 1,4-diketones predominantly. However, the addition of $AlCl_3$ in the reaction mixture at 60°C with increasing the reaction time up to 8 hours allowed the cyclization reaction and selective formation of tetrasubstituted furans with good yields. Obviously, for 1,4-diones containing aromatic, cycloalkyl, haloalkyl or heterocycloalkyl substituents, the use of a stronger

Lewis acid, for example AlCl₃, along with EtAlCl₂, favors the Paal-Knorr cyclization. The reaction goes "to completion" giving the corresponding furans as the end products.

Thus, we have first carried out the reaction between unsymmetrical acetylenes, bearing alkyl, cycloalkyl, heterocycloalkyl, haloalkyl, and aromatic substituents, carboxylic acid esters (alkyl carboxylates, methyl cyclopropanecarboxylate), and EtAlCl₂ catalyzed by Cp₂TiCl₂, and studied the effect of unsymmetrical acetylene substituent on the chemoselectivity of the reaction. It was shown that dialkylsubstituted unsymmetrical acetylenes in these reactions selectively produce substituted furans in ~ 70-75% yield, while unsymmetrical acetylenes with aromatic, cycloalkyl, heterocycloalkyl or haloalkyl substituents enter into the reaction with carboxylic acid esters and EtAlCl₂ to afford individual tetrasubstituted furans (63 – 82%), when using a stronger Lewis acid (AlCl₃) together with EtAlCl₂.

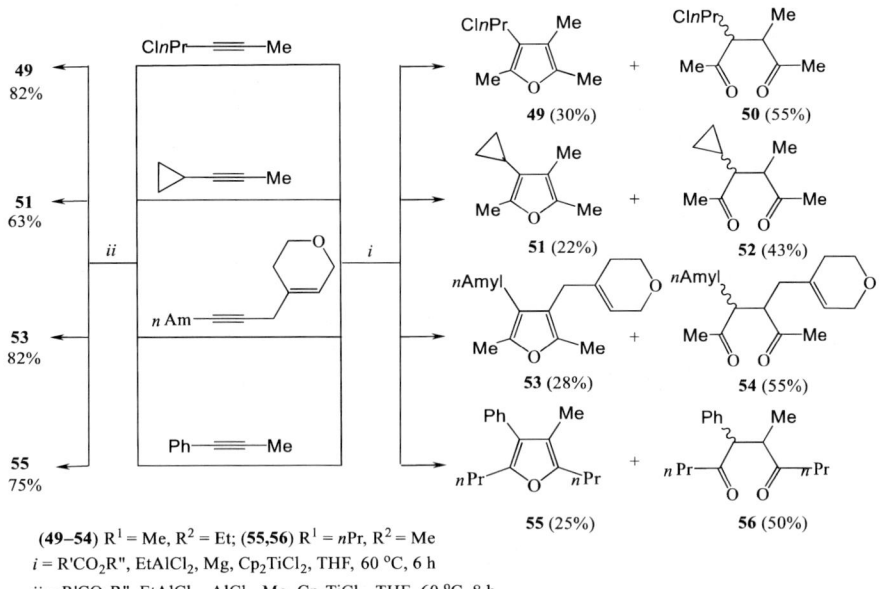

(49–54) R¹ = Me, R² = Et; (55,56) R¹ = nPr, R² = Me
i = R'CO₂R", EtAlCl₂, Mg, Cp₂TiCl₂, THF, 60 °C, 6 h
ii = R'CO₂R", EtAlCl₂, AlCl₃, Mg, Cp₂TiCl₂, THF, 60 °C, 8 h

Scheme 18.

Conclusion

In summary, we have developed a new efficient one-pot procedure for the synthesis of tetraalkyl substituted furans through the reaction of symmetrical acetylenes with alkyl(cycloalkyl, aryl, hetaryl)carboxylic acid esters and esters of α,ω-dicarboxylic acids in the presence of $EtAlCl_2$ under the action of bis(cyclopentadienyl)titanium dichloride as the catalyst.

It was shown that the length of the hydrocarbon linker between the dicarboxylate carbonyl groups significantly affects the reaction pathway.

Symmetrical acetylenes in the reaction with diesters of α,ω-dicarboxylic acids (e.g., pimelic, cork, and azelaic), in which the hydrocarbon chain between the carboxyl groups exceeds four CH_2 units, form tetrasubstituted furans. Diesters, in which the hydrocarbon chain between the carboxyl groups equal to three or four CH_2 units, produce alkylidene and alkenyl substituted C_5–C_6 cyclic ketones.

Unsymmetrical acetylenes with aromatic, cycloalkyl, heterocycloalkyl or haloalkyl substituents enter into the reaction with monocarboxylic acid esters and $EtAlCl_2$ to afford individual tetrasubstituted furans, when using a stronger Lewis acid ($AlCl_3$) together with $EtAlCl_2$.

The developed catalytic method for the synthesis of tetrasubstituted furans allows a variety of previously inaccessible and practically important compounds of this series in one preparative step with good yields from symmetrical and unsymmetrical acetylenes, including functionally substituted, and esters of monocarboxylic and dicarboxylic acids in the presence of $EtAlCl_2$, metallic magnesium, and the Cp_2TiCl_2 catalyst.

References

[1] Lin, J., Liu, S., Sun, B., Niu, S., Li, E., Liu, X. & Che, Y. (2010). Polyketides from the Ascomycete Fungus *Leptosphaeria* sp. *J. Nat. Prod.*, *73*, 905–910.

[2] Rodríguez, B., de la Torre, M. C., Jimeno, M. L., Bruno, M., Fazio, C., Piozzi, F., Savona, G. & Perales, A. (1995). Rearranged neo-clerodane diterpenoids from Teucrium brevifolium and their biogenetic pathway. *Tetrahedron.*, *51*, 837–848.

[3] Coumar, M. S., Chu, C. Y., Lin, C. W., Shiao, H. Y., Ho, Y. L., Reddy, R., Lin, W. H., Chen, C. H., Peng, Y. H., Leou, J. S., Lien, T. W., Huang, C. T., Fang, M. Y., Wu, S. H., Wu, J. S., Chittimalla, S. K., Song, J. S., Hsu, J. T. A., Wu, S. Y., Liao, C. C., Chao, Y. S. & Hsieh, H. P. (2010). Fast-Forwarding Hit to Lead: Aurora and Epidermal Growth Factor Receptor Kinase Inhibitor Lead Identification. *J. Med. Chem.*, *53*, 4980–4988.

[4] Flynn, B. L., Hamel, E. & Jung, M. K. (2002). One-Pot Synthesis of Benzo[*b*]furan and Indole Inhibitors of Tubulin Polymerization. *J. Med. Chem.*, *45*, 2670–2673.

[5] Mortensen, D. S., Rodriguez, A. L., Carlson, K. E., Sun, J., Katzenellenbogen, B. S. & Katzenellenbogen, J. A. (2001). Synthesis and Biological Evaluation of a Novel Series of Furans: Ligands Selective for Estrogen Receptor α. *J. Med. Chem.*, *44*, 3838–3848.

[6] Rahmathullah, S. M., Hall, J. E., Bender, B. C., McCurdy, D. R., Tidwell, R. R. & Boykin, D. W. (1999). Prodrugs for Amidines: Synthesis and Anti-Pneumocystis carinii Activity of Carbamates of 2,5-Bis(4-amidinophenyl)furan. *J. Med. Chem.*, *42*, 3994–4000.

[7] Francesconi, I., Wilson, W. D., Tanious, F. A., Hall, J. E., Bender, B. C., Tidwell, R. R., McCurdy, D. & Boykin, D. W. (1999). 2,4-Diphenyl Furan Diamidines as Novel Anti- *Pneumocystis carinii* Pneumonia Agents. *J. Med. Chem.*, *42*, 2260–2265.

[8] Bai, H. T., Lin, H. C. & Luh, T. Y. (2010). Phenanthrene-Tethered Furan-Containing Cyclophenes: Synthesis and Photophysical Properties. *J. Org. Chem.*, *75*, 4591–4595.

[9] Yi, C., Blum, C., Lehmann, M., Keller, S., Liu, S. X., Frei, G., Neels, A., Hauser, J., Schürch, S. & Decurtins, S. (2010). Versatile Strategy To Access Fully Functionalized Benzodifurans: Redox-Active Chromophores for the Construction of Extended π-Conjugated Materials. *J. Org. Chem.*, *75*, 3350–3357.

[10] Singh, R. P., Foxman, B. M. & Deng, L. (2010). Asymmetric Vinylogous Aldol Reaction of Silyloxy Furans with a Chiral Organic Salt. *J. Am. Chem. Soc.*, *132*, 9558–9560.

[11] Ouairy, C., Michel, P., Delpech, B., Crich, D. & Marazano, C. (2010). Synthesis of *N* -Acyl-5-aminopenta-2,4-dienals via Base-Induced Ring-Opening of *N* -Acylated Furfurylamines: Scope and Limitations. *J. Org. Chem.*, *75*, 4311–4314.

[12] Guindon, Y., Therien, M., Girard, Y. & Yoakim, C. (1987). Regiocontrolled opening of cyclic ethers using dimethylboron bromide. *J. Org. Chem.*, *52*, 1680–1686.

[13] Lipshutz, B. H. (1986). Five-membered heteroaromatic rings as intermediates in organic synthesis. *Chem. Rev.*, *86*, 795–819.

[14] Goldsmith, D. J., Kennedy, E. & Campbell, R. G. (1975). Cleavage of cyclic ethers by magnesium bromide-acetic anhydride. SN2 substitution at a secondary site. *J. Org. Chem.*, *40*, 3571–3574.

[15] Benassi, R. (1996). 2.05 – Furans and their Benzo Derivatives: Structure. in *Comprehensive Heterocyclic Chemistry II*, pp. 259–29.

[16] Dean, F. M. & Sargent, M. V. (1984). 3.10 – Furans and their Benzo Derivatives: (i) Structure. in *Comprehensive Heterocyclic Chemistry*, pp. 531–597.

[17] Paal, C. (1884). Ueber die Derivate des Acetophenonacetessigesters und des Acetonylacetessigesters. *Berichte der Dtsch. Chem. Gesellschaft.*, *17*, 2756–2767. [About the derivatives of Acetophenonacetessigesters and acetonylacetessigesters. *Reports of Dtsch. Chem. Society*.]

[18] Knorr, L. (1884). Synthese von Furfuranderivaten aus dem Diacetbernsteinsäureester. *Berichte der Dtsch. Chem. Gesellschaft.*, *17*, 2863–2870. [Synthesis of Furfural Derivatives from Diacetic Acid Ester. *Reports of Dtsch. Chem. Society*]

[19] Zhang, Y., Chen, Z., Xiao, Y. & Zhang, J. (2009). Rh^{1} -Catalyzed Regio- and Stereospecific Carbonylation of 1-(1-Alkynyl) cyclopropyl Ketones: A Modular Entry to Highly Substituted 5,6-Dihydrocyclopenta[*c*]furan-4-ones. *Chem. - A Eur. J.*, *15*, 5208–5211.

[20] Liu, R. & Zhang, J. (2009). Tetrasubstituted Furans by Pd[II] - Catalyzed Three-Component Domino Reactions of 2-(1-Alkynyl)-2-alken-1-ones with Nucleophiles and Vinyl Ketones or Acrolein. *Chem. - A Eur. J.*, *15*, 9303–9306.

[21] Xiao, Y. & Zhang, J. (2009). Furans versus 4H-pyrans: catalyst-controlled regiodivergent tandem Michael addition–cyclization reaction of 2-(1-alkynyl)-2-alken-1-ones with 1,3-dicarbonyl compounds. *Chem. Commun.*, 3594-3596.

[22] Xiao, Y. & Zhang, J. (2008). Tetrasubstituted Furans by a PdII-Catalyzed Three-Component Michael Addition/Cyclization/Cross-Coupling Reaction. *Angew. Chemie Int. Ed.*, *47*, 1903–1906.

[23] Zhang, J. & Schmalz, H. G. (2006). Gold(I)-Catalyzed Reaction of 1-(1-Alkynyl)-cyclopropyl Ketones with Nucleophiles: A Modular Entry to Highly Substituted Furans. *Angew. Chemie Int. Ed.*, *45*, 6704–6707.

[24] Yao, T., Zhang, X. & Larock, R. C. (2004). AuCl$_3$ -Catalyzed Synthesis of Highly Substituted Furans from 2-(1-Alkynyl)-2-alken-1-ones. *J. Am. Chem. Soc.*, *126*, 11164–11165.

[25] Sniady, A., Durham, A., Morreale, M. S., Wheeler, K. A. & Dembinski, R. (2007). Room Temperature Zinc Chloride-Catalyzed Cycloisomerization of Alk-3-yn-1-ones: Synthesis of Substituted Furans. *Org. Lett.*, *9*, 1175–1178.

[26] Imagawa, H., Kurisaki, T. & Nishizawa, M. (2004). Mercuric Triflate-Catalyzed Synthesis of 2-Methylfurans from 1-Alkyn-5-ones. *Org. Lett.*, *6*, 3679–3681.

[27] Patil, N. T., Wu, H. & Yamamoto, Y. (2005). Cu(I) Catalyst in DMF: An Efficient Catalytic System for the Synthesis of Furans from 2-(1-Alkynyl)-2-alken-1-ones. *J. Org. Chem.*, *70*, 4531–4534.

[28] Zhao, L. B., Guan, Z. H., Han, Y., Xie, Y. X., He, S. & Liang, Y. M. (2007). Copper-Catalyzed [4 + 1] Cycloadditions of α,β-Acetylenic Ketones with Diazoacetates to Form Trisubstituted Furans. *J. Org. Chem.*, *72*, 10276–10278.

[29] Li, Y. & Yu, Z. (2009). Palladium-Catalyzed Carbonylative Cycloisomerization of γ-Propynyl-1,3-diketones: A Concise Route to Polysubstituted Furans. *J. Org. Chem.*, *74*, 8904–8907.

[30] Ma, S. & Yu, Z. (2002). Oxidative Cyclization-Dimerization Reaction of 2,3-Allenoic Acids and 1,2-Allenyl Ketones: An Efficient Synthesis of 4-(3'-Furanyl)butenolide Derivatives. *Angew. Chemie Int. Ed.*, *41*, 1775–1778.

[31] Peng, L., Zhang, X., Ma, M. & Wang, J. (2007). Transition-Metal-Catalyzed Rearrangement of Allenyl Sulfides: A Route to Furan Derivatives. *Angew. Chemie Int. Ed.*, *46*, 1905–1908.

[32] Hashmi, A. S. K., Schwarz, L., Choi, J. H. & Frost, T. M. (2000). A New Gold-Catalyzed C—C Bond Formation. *Angew. Chemie Int. Ed.*, *39*, 2285–2288.

[33] Dudnik, A. S. & Gevorgyan, V. (2007). Metal-Catalyzed [1,2]-Alkyl Shift in Allenyl Ketones: Synthesis of Multisubstituted Furans. *Angew. Chemie Int. Ed.*, *46*, 5195–5197.

[34] Ma, S. & Zhang, J. (2000). Pd0-catalyzed cyclization reaction of aryl or alk-1-enyl halides with 1,2-dienyl ketones: a general and efficient synthesis of polysubstituted furans. *Chem. Commun.*, 117–118.

[35] Sromek, A. W., Rubina, M. & Gevorgyan, V. (2005). 1,2-Halogen Migration in Haloallenyl Ketones: Regiodivergent Synthesis of Halofurans. *J. Am. Chem. Soc.*, *127*, 10500–10501.

[36] Dudnik, A. S., Xia, Y., Li, Y. & Gevorgyan, V. (2010). Computation-Guided Development of Au-Catalyzed Cycloisomerizations Proceeding via 1,2-Si or 1,2-H Migrations: Regiodivergent Synthesis of Silylfurans. *J. Am. Chem. Soc.*, *132*, 7645–7655.

[37] Zhou, C. Y., Chan, P. W. H. & Che, C. M. (2006). Gold(III) Porphyrin-Catalyzed Cycloisomerization of Allenones. *Org. Lett.*, *8*, 325–328.

[38] Ma, S. & Li, L. (2000). Palladium-Catalyzed Cyclization Reaction of Allylic Bromides with 1,2-Dienyl Ketones. An Efficient Synthesis of 3-Allylic Polysubstituted Furans. *Org. Lett.*, *2*, 941–944.

[39] Ma, S., Lu, L. & Zhang, J. (2004). Catalytic Regioselectivity Control in Ring-Opening Cycloisomerization of Methylene- or Alkylidenecyclopropyl Ketones. *J. Am. Chem. Soc.*, *126*, 9645–9660.

[40] Ma, S. & Zhang, J. (2003). 2,3,4- or 2,3,5-Trisubstituted Furans: Catalyst-Controlled Highly Regioselective Ring-Opening Cycloisomerization Reaction of Cyclopropenyl Ketones. *J. Am. Chem. Soc.*, *125*, 12386–12387.

[41] Padwa, A., Kassir, J. M. & Xu, S. L. (1991). Rhodium-catalyzed ring-opening reaction of cyclopropenes. Control of regioselectivity by the oxidation state of the metal. *J. Org. Chem.*, *56*, 6971–6972.

[42] Hashmi, A. S. K. & Sinha, P. (2004). Gold Catalysis: Mild Conditions for the Transformation of Alkynyl Epoxides to Furans. *Adv. Synth. Catal.*, *346*, 432–438.

[43] Marshall, J. A. & DuBay, W. J. (1992). Base-catalyzed isomerization of alkynyloxiranes. A general synthesis of furans. *J. Am. Chem. Soc.*, *114*, 1450–1456.

[44] Aurrecoechea, J. M., Pérez, E. & Solay, M. (2001). Synthesis of Trisubstituted Furans from Epoxypropargyl Esters by Sequential SmI$_2$-Promoted Reduction−Elimination and Pd(II)-Catalyzed Cycloisomerization. *J. Org. Chem.*, *66*, 564–569.

[45] Lo, C. Y., Guo, H., Lian, J. J., Shen, F. M. & Liu, R. S. (2002). Efficient Synthesis of Functionalized Furans via Ruthenium-Catalyzed Cyclization of Epoxyalkyne Derivatives. *J. Org. Chem.*, *67*, 3930–3932.

[46] Blanc, A., Tenbrink, K., Weibel, J. M. & Pale, P. (2009). Silver(I)-Catalyzed Cascade: Direct Access to Furans from Alkynyloxiranes. *J. Org. Chem.*, *74*, 4360–4363.

[47] Aurrecoechea, J. M., Durana, A. & Pérez, E. (2008). Palladium-Catalyzed Cyclization/Heck- and Cyclization/Conjugate-Addition-Type Sequences in the Preparation of Polysubstituted Furans. *J. Org. Chem.*, *73*, 3650–3653.

[48] Nishibayashi, Y., Yoshikawa, M., Inada, Y., Milton, M. D., Hidai, M. & Uemura, S. (2003) Novel Ruthenium- and Platinum-Catalyzed Sequential Reactions: Synthesis of Tri- and Tetrasubstituted Furans

and Pyrroles from Propargylic Alcohols and Ketones. *Angew. Chemie Int. Ed.*, *42*, 2681–2684.

[49] Arimitsu, S. & Hammond, G. B. (2007). Selective Synthesis of Fluorinated Furan Derivatives via $AgNO_3$ -Catalyzed Activation of an Electronically Deficient Triple Bond. *J. Org. Chem.*, *72*, 8559–8561.

[50] Nanayakkara, P. & Alper, H. (2006). Synthesis of 3-Substituted Furans by Hydroformylation. *Adv. Synth. Catal.*, *348*, 545–550.

[51] Pan, Y., Zhao, S., Ji, W. & Zhan, Z. (2009). One-Pot Synthesis of Substituted Furans Using $Cu(OTf)_2$-Catalyzed Propargylation/ Cycloisomerization Tandem Reaction. *J. Comb. Chem.*, *11*, 103–109.

[52] Egi, M., Azechi, K. & Akai, S. (2009). Cationic Gold(I)-Mediated Intramolecular Cyclization of 3-Alkyne-1,2-diols and 1-Amino-3-alkyn-2-ols: A Practical Route to Furans and Pyrroles. *Org. Lett.*, *11*, 5002–5005.

[53] Aponick, A., Li, C. Y., Malinge, J. & Marques, E. F. (2009). An Extremely Facile Synthesis of Furans, Pyrroles, and Thiophenes by the Dehydrative Cyclization of Propargyl Alcohols. *Org. Lett.*, *11*, 4624–4627.

[54] Du, X., Chen, H. & Liu, Y. (2008). New Synthetic Approach for the Construction of Multisubstituted 2-Acyl Furans by the IBX-Mediated Cascade Oxidation/Cyclization of *cis*-2-En-4-yn-1-ols (IBX=2-Iodoxybenzoic Acid). *Chem. - A Eur. J.*, *14*, 9495–9498.

[55] Zhang, X., Lu, Z., Fu, C. & Ma, S. (2010). Synthesis of Polysubstituted Furans Based on a Stepwise Sonogashira Coupling of (Z)-3-Iodoalk-2-en-1-ols with Terminal Propargylic Alcohols and Subsequent Au(I)- or Pd(II)-Catalyzed Cyclization−Aromatization via Elimination of H_2O. *J. Org. Chem.*, *75*, 2589–2598.

[56] Gabriele, B., Salerno, G. & Lauria, E. (1999). A General and Facile Synthesis of Substituted Furans by Palladium-Catalyzed Cycloisomerization of (Z)-2-En-4-yn-1-ols. *J. Org. Chem.*, *64*, 7687–7692.

[57] Liu, Y., Song, F., Song, Z., Liu, M. & Yan, B. (2005). Gold-Catalyzed Cyclization of (Z)-2-En-4-yn-1-ols: Highly Efficient

Synthesis of Fully Substituted Dihydrofurans and Furans. *Org. Lett.*, *7*, 5409–5412.

[58] Schneider, C. C., Caldeira, H., Gay, B. M., Back, D. F. & Zeni, G. (2010). Transmetalation of Z-Telluroenynes: Stereoselective Synthesis of Z-Enynols and Their Application in Palladium-Catalyzed Cyclization. *Org. Lett.*, *12*, 936–939.

[59] Jiang, H., Yao, W., Cao, H., Huang, H. & Cao, D. (2010). Iron-Catalyzed Domino Process for the Synthesis of α-Carbonyl Furan Derivatives via One-Pot Cyclization Reaction. *J. Org. Chem.*, *75*, 5347–5350.

[60] Suhre, M. H., Reif, M. & Kirsch, S. F. (2005). Gold(I)-Catalyzed Synthesis of Highly Substituted Furans. *Org. Lett.*, *7*, 3925–3927.

[61] Cao, H., Jiang, H., Yao, W. & Liu, X. (2009). Copper-Catalyzed Domino Rearrangement/Dehydrogenation Oxidation/Carbene Oxidation for One-Pot Regiospecific Synthesis of Highly Functionalized Polysubstituted Furans. *Org. Lett.*, *11*, 1931–1933.

[62] Cao, H., Jiang, H., Yuan, G., Chen, Z., Qi, C. & Huang, H. (2010). Nano-Cu2O-Catalyzed Formation of C–C and C–O Bonds: One-Pot Domino Process for Regioselective Synthesis of α-Carbonyl Furans from Electron-Deficient Alkynes and 2-Yn-1-ols. *Chem. - A Eur. J.*, *16*, 10553–10559.

[63] Cao, H., Jiang, H., Mai, R., Zhu, S. & Qi, C. (2010). Silver-Catalyzed One-Pot Cyclization Reaction of Electron- Deficient Alkynes and 2-Yn-1-ols: An Efficient Domino Process to Polysubstituted Furans. *Adv. Synth. Catal.*, *352*, 143–152.

[64] Zhang, M., Jiang, H. F., Neumann, H., Beller, M. & Dixneuf, P. H. (2009). Sequential Synthesis of Furans from Alkynes: Successive Ruthenium(II)- and Copper(II)-Catalyzed Processes. *Angew. Chemie Int. Ed.*, *48*, 1681–1684.

[65] Barluenga, J., Riesgo, L., Vicente, R., López, L. A. & Tomás, M. (2008). Cu(I)-Catalyzed Regioselective Synthesis of Polysubstituted Furans from Propargylic Esters via Postulated (2-Furyl)carbene Complexes. *J. Am. Chem. Soc.*, *130*, 13528–13529.

[66] Xu, B. & Hammond, G. B. (2006). A New Convenient Synthesis of Propargylic Fluorohydrins and 2,5-Disubstituted Furans from Fluoropropargyl Chloride. *J. Org. Chem.*, *71*, 3518–3521.

[67] Xu, L., Huang, X. & Zhong, F. (2006). Intermolecular Tandem Addition−Cyclization of Bromoallenes: A Facile Synthesis of Methylenecyclopropyl Carboxylates and Polysubstituted Furans. *Org. Lett.*, *8*, 5061–5064.

[68] Donohoe, T. J., Fishlock, L. P., Lacy, A. R. & Procopiou, P. A. (2007). A Metathesis-Based Approach to the Synthesis of Furans. *Org. Lett.*, *9*, 953–956.

[69] Liu, W., Jiang, H., Zhang, M. & Qi, C. (2010). Synthetic Approach to Polysubstituted Furans: An Efficient Addition/Oxidative Cyclization of Alkynoates and 1,3-Dicarbonyl Compounds. *J. Org. Chem.*, *75*, 966–968.

[70] Li, H. & Hsung, R. P. (2009). Highly Substituted 2-Amido-furans From Rh(II)-Catalyzed Cyclopropenations of Ynamides. *Org. Lett.*, *11*, 4462–4465.

[71] Sydnes, L. K., Holmelid, B., Sengee, M. & Hanstein, M. (2009). New Regiospecific Synthesis of Tri- and Tetra-Substituted Furans. *J. Org. Chem.*, *74*, 3430–3443.

[72] Yang, Y. K., Choi, J. H. & Tae, J. (2005). Synthesis of 2,3-Di- and 2,3,4-Trisubstituted Furans from 1,2-Dioxines Generated by an Enyne-RCM/Diels−Alder Reaction Sequence. *J. Org. Chem.*, *70*, 6995–6998.

[73] Dzhemilev, U. M., Khafizova, L. O., Gubaidullin, R. R., Khalilov, L. M. & Ibragimov, A. G. (2009). The first one-pot synthesis of alkoxycyclopropanes via cyclometalation of styrene with Cl_nAlEt_{3-n} and RCO_2R' mediated by Cp_2ZrCl_2. *Tetrahedron Lett.*, *50*, 7086–7088.

[74] Khafizova, L. O., Gubaidullin, R. R. & Dzhemilev, U. M. (2011). Zirconium-catalyzed cyclopropanation of α-olefins mediated by $R'CO_2R''$ and Cl_nAlEt_{3-n}. *Tetrahedron.*, *67*, 9142–9147.

[75] Sato, F. & Urabe, H. Titanium(II) Alkoxides in Organic Synthesis. in *Titanium and Zirconium in Organic Synthesis*, pp. 319–354, Wiley-

VCH Verlag GmbH & Co. KGaA, Weinheim, FRG, 10.1002/ 3527600671.ch9.

[76] Sato, F., Urabe, H. & Okamoto, S. (2000). Synthesis of Organotitanium Complexes from Alkenes and Alkynes and Their Synthetic Applications. *Chem. Rev.*, *100*, 2835–2886.

[77] Eisch, J. J. (2001). Early transition metal carbenoid reagents in epimetallation and metallative dimerization of unsaturated organic substrates. *J. Organomet. Chem.*, *617*, 148–157.

[78] Kulinkovich, O. G. & de Meijere, A. (2000). 1, *n*-Dicarbanionic Titanium Intermediates from Monocarbanionic Organometallics and Their Application in Organic Synthesis. *Chem. Rev.*, *100*, 2789–2834.

[79] Wu, X. F. & Wu, X. F. (2016). Chapter 2 – Synthesized by [2+2+1] Cyclization Reactions. in *Transition Metal Catalyzed Furans Synthesis*, pp. 3–7, 10.1016/B978-0-12-804034-8.00002-7.

[80] Yin, G., Wang, Z., Chen, A., Gao, M., Wu, A. & Pan, Y. (2008). A New Facile Approach to the Synthesis of 3-Methylthio-Substituted Furans, Pyrroles, Thiophenes, and Related Derivatives. *J. Org. Chem.*, *73*, 3377–3383.

[81] Haddadin, M. J., Agha, B. J. & Tabri, R. F. (1979). Syntheses of some furans and naphtho[2,3-c]derivatives of furan, pyrrole and thiophene. *J. Org. Chem.*, *44*, 494–497.

[82] Dzhemilev, U. M., Ibragimov, A. G., Khafizova, L. O., Yakupova, L. R. & Khalilov, L. M. (2005). Cycloaluminizing of Acetylenes and 1,4-Enynes in the Presence of Zr-containing Catalysts. *Russ. J. Org. Chem.*, *41*, 667–672.

[83] Szakal-Quin, G., Graham, D. G., Millington, D. S., Maltby, D. A. & McPhail, A. T. (1986). Stereoisomer effects on the Paal-Knorr synthesis of pyrroles. *J. Org. Chem.*, *51*, 621–624.

[84] Shaibakova, M. G., Khafizova, L. O., Chobanov, N. M., Gubaidullin, R. R., Popod'ko, N. R. & Dzhemilev, U. M. (2014). The efficient one-pot synthesis of tetraalkyl substituted furans from symmetrical acetylenes, EtAlCl2, and carboxylic esters catalyzed by Cp$_2$TiCl$_2$. *Tetrahedron Lett.*, *55*, 1326–1328.

[85] Khafizova, L. O., Shaibakova, M. G., Chobanov, N. M., Tyumkina, T. V., Gubaidullin, R. R., Popod'ko, N. R. & Dzhemilev, U. M. (2016). Effective one-pot synthesis of 2,3-dialkyl-1,4-dicyclopropyl-butane-1,4-diones catalyzed by Cp2TiCl2. *Mendeleev Commun.*, *26*, 223–225.

[86] Mothana, B. & Boyd, R. J. (2007). A density functional theory study of the mechanism of the Paal–Knorr pyrrole synthesis. *J. Mol. Struct. THEOCHEM.*, *811*, 97–107.

[87] Amarnath, V., Amarnath, K., Amarnath, K., Davies, S. & Roberts, L. J. (2004). Pyridoxamine: An Extremely Potent Scavenger of 1,4-Dicarbonyls. *Chem. Res. Toxicol.*, *17*, 410–415.

[88] Amarnath, V., Anthony, D. C., Amarnath, K., Valentine, W. M., Wetterau, L. A. & Graham, D. G. (1991). Intermediates in the Paal-Knorr synthesis of pyrroles. *J. Org. Chem.*, *56*, 6924–6931.

[89] Amarnath, V. & Amarnath, K. (1995). Intermediates in the Paal-Knorr Synthesis of Furans. *J. Org. Chem.*, *60*, 301–307.

[90] Wu, Y. D. & Yu, Z. X. (2001). A Theoretical Study on the Mechanism and Diastereoselectivity of the Kulinkovich Hydroxycyclopropanation Reaction. *J. Am. Chem. Soc.*, *123*, 5777–5786.

[91] Kulinkovich, O. G. (2000). Titanacyclopropanes as versatile intermediates for carbon-carbon bond formation in reactions with unsaturated compounds. *Pure Appl. Chem.*, *72*, 1715–1719.

[92] Dewar, M. J. S. & Harris, J. M. (1968). Rates of solvolysis of 2-cyclopropylethyl brosylates. *J. Am. Chem. Soc.*, *90*, 4468–4469.

[93] Nikoletić, M., Borčić, S. & Sunko, D. E. (1967). Secondary hydrogen isotope effects—IX: Solvolysis rates of methyl and methyl-d3 substituted cyclopropylcarbinyl and cyclobutyl derivatives. *Tetrahedron.*, *23*, 649–660.

[94] Lambert, J. B., Napoli, J. J., Johnson, K. K. & Taba, K. N. (1985). Scope, limitations, and mechanism of the homoconjugate electrophilic addition of hydrogen halides. *J. Org. Chem.*, *50*, 1291–1295.

[95] Frost, J. R., Cheong, C. B., Akhtar, W. M., Caputo, D. F. J., Stevenson, N. G. & Donohoe, T. J. (2015). Strategic Application and Transformation of *ortho*-Disubstituted Phenyl and Cyclopropyl Ketones To Expand the Scope of Hydrogen Borrowing Catalysis. *J. Am. Chem. Soc.*, *137*, 15664–15667.

[96] Kagermeier, N., Werner, K., Keller, M., Baumeister, P., Bernhardt, G., Seifert, R. & Buschauer, A. (2015). Dimeric carbamoyl-guanidine-type histamine H2 receptor ligands: A new class of potent and selective agonists. *Bioorg. Med. Chem.*, *23*, 3957–3969.

[97] Yang, T. P., Li, Q., Lin, J. H. & Xiao, J. C. (2014). Boron-trihalide-promoted regioselective ring-opening reactions of gem-difluorocyclopropyl ketones. *Chem. Commun.*, *50*, 1077–1079.

[98] Srikrishna, A. & Reddy, T. J. (2001). Chiral synthons from carvone. Part 50.† Enantiospecific approaches to both enantiomers of bicyclo[4.3.0]nonane-3,8-dione derivatives. *J. Chem. Soc. Perkin Trans.*, *1*. 10.1039/b104253j.

[99] Déziel, R., Malenfant, E., Thibault, C., Fréchette, S. & Gravel, M. (1997). 2,6-Bis[(2S)-tetrahydrofuran-2-yl]phenyl diselenide: An effective reagent for asymmetric electrophilic addition reactions to olefins. *Tetrahedron Lett.*, *38*, 4753–4756.

[100] Grieco, P. A. & Masaki, Y. (1975). Synthesis of the Valeriana waalichi hydrocarbon sesquifenchene. Route to specifically functionalized 7,7-disubstituted bicyclo[2.2.1]heptane derivatives. *J. Org. Chem.*, *40*, 150–151.

[101] Khafizova, L. O., Shaibakova, M. G., Chobanov, N. M., Gubaidullin, R. R., Tyumkina, T. V. & Dzhemilev, U. M. (2015). Synthesis of tetrasubstituted furans by multicomponent reaction of alkynes with dichloro(ethyl)aluminum and carboxylic acid esters in the presence of Cp_2TiCl_2. *Russ. J. Org. Chem.*, *51*, 1277–1281.

[102] Shaibakova, M. G., Khafizova, L. O., Chobanov, N. M., Gubaidullin, R. R., Popod'ko, N. R. & Dzhemilev, U. M. (2014). The efficient one-pot synthesis of tetraalkyl substituted furans from symmetrical

acetylenes, EtAlCl$_2$, and carboxylic esters catalyzed by Cp$_2$TiCl$_2$. *Tetrahedron Lett.*, *55*, 1326–1328.

[103] Khafizova, L. O., Shaibakova, M. G., Chobanov, N. M., Tyumkina, T. V., Gubaidullin, R. R., Popod'ko, N. R. & Dzhemilev, U. M. (2016). Effective one-pot synthesis of 2,3-dialkyl-1,4-dicyclopropyl-butane-1,4-diones catalyzed by Cp$_2$TiCl$_2$. *Mendeleev Commun.*, *26*, 223–225.

INDEX

#

5-exo-dig cyclization, vii, viii, 68, 75, 77, 78, 79, 80

A

acetic acid, 22, 100, 105, 112
acetone, 7, 8, 9, 39, 41
acid, ix, 2, 4, 5, 6, 8, 9, 11, 13, 14, 15, 17, 18, 19, 20, 21, 22, 24, 25, 26, 27, 28, 29, 30, 31, 32, 34, 36, 37, 38, 42, 50, 53, 54, 55, 70, 71, 73, 76, 77, 81, 92, 93, 99, 100, 102, 105, 106, 108, 109, 110, 111, 113, 114
acidic, 12, 27, 28, 29, 38, 39
activation parameters, 19, 20
active compound, 70
active site, 97
alcohols, 8, 9, 11, 13, 20, 25, 35, 69, 75, 76, 78, 82
aldehydes, 14, 15, 16, 18, 19, 20, 21, 22, 24, 25, 28, 29, 34, 42, 48, 49, 50, 51, 58, 62
alkylation, 13, 70

alkynes, 68, 80, 81, 84, 85, 86, 89, 121, 123, 125
amines, 35, 36, 103
ammonia, 36
ammonium, 36
antitumor, viii, 68, 70
aqueous solutions, 44, 52
aromatic compounds, vii, viii, 1
atoms, 97
autocatalysis, 31, 32

B

base catalysis, 19
basic research, 57, 61, 64
benzene, 2, 15, 22, 25, 26, 28, 53, 96
benzopyrazine, 69
bioavailability, viii, 68, 69
biological activities, 92
biologically active compounds, 68, 69, 92
bonding, 23, 24
bonds, 25, 27, 93, 108
bromine, 2, 44
building blocks, 92
by-products, 42, 94

Index

C

carbon, 2, 5, 22, 74, 75, 98, 107, 124
carbon atoms, 74, 75, 98
carbon dioxide, 2
carbon monoxide, 2
carbonyl groups, 109, 114
carboxyl, ix, 92, 109, 110, 114
carboxylic acid, vii, ix, 13, 15, 16, 24, 35, 91, 93, 98, 99, 100, 101, 102, 110, 111, 112, 113, 114, 125
carboxylic acid esters, vii, ix, 91, 93, 98, 99, 101, 102, 111, 113, 114, 125
catalysis, v, 5, 19, 46, 50, 52, 59, 63, 67, 76, 80, 83, 87, 89, 91, 92, 119, 125
catalyst, vii, ix, 7, 8, 9, 10, 11, 12, 13, 27, 31, 32, 35, 36, 38, 39, 40, 41, 42, 46, 48, 55, 68, 77, 78, 80, 82, 91, 93, 94, 95, 96, 97, 99, 101, 102, 104, 106, 108, 110, 112, 114, 117
catalytic activity, 78, 95
catalytic system, viii, 68, 76
chemical, vii, 1, 15, 32, 42, 44, 46, 62, 98, 103, 105
chloride anion, 81
chlorine, 81
chloroform, 36, 40, 41
chromatography, 9, 11, 13, 30, 41, 81, 102
classical methods, 69
composition, 9, 10, 13, 14, 16, 18, 20, 31, 32, 34
compounds, viii, 2, 4, 5, 6, 7, 8, 9, 10, 13, 14, 15, 16, 19, 27, 28, 30, 31, 32, 34, 36, 37, 39, 40, 41, 42, 43, 45, 46, 47, 48, 49, 50, 51, 52, 53, 54, 56, 57, 58, 59, 61, 63, 64, 67, 69, 70, 71, 72, 74, 75, 77, 78, 80, 81, 92, 103, 107, 108, 114, 117, 124
conjugation, 15, 22, 27, 33, 38, 101
consumption, 12, 15, 34, 35, 36, 38
coordination, 77, 82
coupling constants, 107

Cp_2TiCl_2, vii, ix, 91, 92, 93, 95, 96, 97, 98, 99, 101, 102, 104, 106, 108, 109, 110, 112, 113, 114, 123, 124, 125, 126
Cp_2TiCl_2 catalyst, vii, ix, 91, 93, 95, 99, 104, 108, 110, 112, 114
crystalline, 21, 36, 39, 40

D

decomposition, viii, 1, 16, 20, 31, 33
density functional theory, 124
derivatives, vii, viii, 2, 10, 40, 41, 42, 45, 50, 51, 53, 54, 58, 62, 64, 68, 69, 71, 72, 73, 76, 77, 83, 92, 100, 102, 123, 124, 125
destruction, 76, 80
dichloroethane, 37, 38
dienes, vii, viii, 1
dimerization, 123
dimethylformamide, 5
distillation, 34, 38, 40, 41, 42

E

education, 57, 61
electron, 14, 15, 18, 20, 22, 24, 25, 27, 33, 34, 46, 80
electron paramagnetic resonance, 46
energy, 4, 103, 104, 107
equilibrium, 28, 29, 39
ester, 14, 20, 26, 28, 29, 32, 72, 78, 93, 94, 95, 96, 99, 100, 102, 106, 108, 110, 111
$EtAlCl_2$, vii, ix, 91, 92, 93, 94, 95, 96, 97, 98, 99, 100, 101, 102, 103, 104, 106, 110, 112, 113, 114, 123, 126
ethanol, 5, 8, 9, 10, 11, 13, 16, 17, 20, 35, 40, 42, 46, 48, 56
ethers, 92, 116
ethyl acetate, 38, 94, 95, 96, 97, 98, 112
ethyl aluminum dichloride, 93, 99, 102, 105, 110

evidence, 98
execution, 4

F

filtration, 35
force, 80
formation, 3, 4, 5, 6, 7, 10, 11, 13, 14, 15, 16, 17, 18, 19, 20, 22, 24, 25, 26, 27, 28, 30, 31, 32, 33, 36, 38, 39, 50, 51, 52, 72, 73, 75, 82, 93, 94, 96, 98, 99, 103, 104, 105, 106, 108, 110, 111, 112, 124
formylfurans, v, vii, 1, 14, 15, 17, 19, 32, 33, 34, 42, 51, 53, 58, 59, 62, 63, 64
fragments, 69
free rotation, 98
furan, vii, viii, 1, 2, 3, 4, 5, 6, 7, 8, 9, 10, 11, 13, 14, 15, 16, 18, 19, 20, 21, 22, 24, 25, 26, 27, 28, 29, 32, 33, 34, 38, 39, 41, 42, 43, 44, 45, 46, 47, 48, 49, 50, 51, 53, 54, 56, 57, 58, 59, 61, 62, 63, 64, 68, 69, 70, 72, 73, 74, 75, 76, 78, 79, 80, 81, 82, 83, 92, 94, 95, 96, 97, 99, 100, 101, 102, 104, 105, 112, 115, 116, 123
furan and hydrofuran compounds, 30, 34, 46, 47, 50, 56
furan-fused triterpenoids, 74, 79, 83
furans, v, vii, viii, 1, 2, 4, 9, 13, 14, 32, 34, 42, 43, 44, 46, 48, 51, 53, 57, 58, 59, 61, 62, 63, 64, 68, 70, 72, 76, 77, 78, 83, 84, 85, 86, 87, 88, 92, 93, 98, 99, 102, 103, 106, 110, 111, 112, 113, 114, 115, 116, 117, 118, 119, 120, 121, 122, 123, 124, 125

G

gold complexes, vii, viii, 68, 76

H

heterocycles, viii, 53, 68, 69, 76, 78, 86, 88, 103
heterocyclization, v, 67, 68, 69, 70, 76, 77, 78, 80
heterogeneous systems, 37
higher education, 57, 60, 63
hydrogen, vii, 1, 4, 5, 6, 7, 8, 9, 10, 11, 14, 15, 16, 18, 19, 20, 22, 24, 25, 26, 27, 28, 29, 32, 33, 34, 35, 36, 37, 38, 39, 40, 41, 42, 44, 45, 46, 47, 48, 49, 50, 51, 52, 53, 54, 56, 57, 58, 61, 62, 63, 64, 124
hydrogen bonds, 25
hydrogen peroxide, v, vii, 1, 2, 4, 5, 6, 7, 8, 9, 10, 11, 14, 15, 16, 18, 19, 20, 22, 25, 26, 27, 28, 29, 32, 33, 34, 35, 36, 37, 38, 39, 40, 41, 42, 44, 45, 46, 47, 48, 49, 50, 51, 52, 53, 54, 56, 57, 58, 61, 62, 63, 64
hydrolysis, 13, 16, 20, 26, 27, 28, 29, 36, 37, 76, 94
hydroperoxides, 5
hydroquinone, 39
hydroxyacids, 31
hydroxyl, 6, 10, 13, 15, 22, 27, 33, 46

I

indole, 69, 115
inventions, 47, 49, 53, 54, 55, 56, 57, 60, 63, 65
ions, ix, 47, 91
IR spectroscopy, 102
isolation, 35, 41, 73
isomerization, 76, 82, 119
isomers, 2, 3, 4, 9, 11, 14, 107
isotope, 124
isoxazole, 69

K

ketones, ix, 21, 22, 69, 70, 75, 76, 78, 82, 83, 92, 108, 110, 114, 118, 125
kinetic studies, 46
kinetics, 20, 103

L

Lewis acids, 76, 82, 102, 105, 110

M

magnesium, vii, ix, 91, 108, 112, 114, 116
mass, 4, 9, 11, 13, 26, 97, 102
mass spectrometry, 4, 9, 11, 13, 26, 97, 102
materials, 92, 103
matter, iv
media, 2, 6, 8, 9, 13, 19, 25, 27, 29, 32, 41, 52, 55
medicinal chemistry, vii, viii, 67, 69, 86, 87, 88, 89
melting, 40, 41
melts, 39
mercury, 28
metabolites, viii, 67, 69
metal complex catalysis, 76
metal complexes, 80
metal oxides, 53
metals, 42
methanol, 4, 42
methodology, 57, 61, 63
methods of synthesis, 34, 57, 61
methyl group, 22, 75, 97
methyl groups, 97
methylene chloride, 96
migration, 22, 25
model system, 30
models, viii, 68
molecular oxygen, 2, 4, 44

molecules, viii, 1, 19, 25, 27, 70, 76, 92
molybdenum, 27, 32, 39, 50
monocarboxylic acid esters, 92, 109, 114

N

Na_2SO_4, 37, 39
niobium, 27, 31, 54
nitrogen, viii, 68, 69
NMR, 3, 25, 26, 74, 75, 97, 102, 107
nonane, 125
nucleus, 10, 13

O

olefins, 93, 122, 125
organic peroxides, 5, 13, 15
oxidation, viii, 1, 2, 3, 4, 5, 6, 7, 8, 9, 10, 11, 12, 13, 14, 15, 16, 17, 18, 19, 20, 21, 22, 24, 25, 26, 27, 28, 30, 31, 32, 34, 35, 36, 37, 38, 39, 40, 41, 42, 43, 44, 45, 46, 47, 48, 49, 50, 51, 54, 55, 56, 58, 62, 64, 119, 120, 121
oxidation products, 9, 11, 12, 13, 15, 16, 18, 24, 42, 48
oxygen, viii, 2, 4, 5, 11, 15, 16, 22, 25, 32, 33, 43, 44, 68, 69, 75, 76, 78, 82
ozonides, 2, 4, 30, 44

P

pathway, 21, 24, 32, 75, 82, 99, 109, 114, 115
pentacyclic triterpenoids, v, vii, viii, 67, 68, 70, 76, 83, 85, 86
peroxide, vii, 1, 4, 5, 6, 7, 8, 9, 10, 11, 14, 15, 16, 18, 19, 20, 21, 22, 24, 25, 26, 27, 28, 31, 32, 33, 34, 35, 36, 37, 38, 39, 40, 41, 42, 44, 45, 46, 47, 48, 49, 50, 51, 52, 53, 54, 56, 57, 58, 61, 62, 63, 64

pH, 7, 11, 17, 24, 25, 28, 29, 34, 36, 38, 39, 42, 52, 55
pharmaceutical, vii, viii, 68
pharmaceuticals, 68, 81
pharmacological research, 69
photochemical transformations, 2
physicochemical methods, 102
polarization, 22, 25
potassium, 70
preparation, iv, ix, 4, 49, 53, 54, 55, 56, 68, 71, 83
protons, 25, 75, 98, 107, 108
pyrazine, 69
pyrazole, 69
pyridine, 15, 20, 21, 25, 43, 69, 79, 80

R

radicals, 6, 10, 13, 15, 22, 32, 46
reaction center, 15
reaction mechanism, 38, 47
reaction medium, 11, 17, 25, 27, 28, 42
reaction order, 19
reaction rate, 12, 19, 20
reaction temperature, 42
reaction time, 9, 36, 37, 72, 95, 97, 106, 110, 112
reactions, vii, viii, 1, 4, 5, 6, 14, 15, 16, 18, 19, 21, 22, 24, 25, 27, 29, 30, 31, 32, 33, 36, 42, 43, 44, 50, 53, 57, 58, 61, 62, 63, 68, 69, 70, 76, 81, 82, 83, 92, 93, 99, 100, 103, 105, 106, 108, 109, 113, 124, 125
reactivity, 4, 13, 100, 101, 112
reagents, vii, 1, 36, 46, 53, 59, 63, 64, 73, 93, 123
regioselectivity, 75, 83, 119
residue, 35, 37, 38, 40, 41
resistance, 60, 63, 65
rings, 26, 69, 77, 116

room temperature, 19, 26, 70, 71, 72, 78, 80, 81, 83

S

scientific papers, 57, 61, 63
seedlings, 60, 63, 65
selectivity, 48, 81
selenium, 27, 52
sensitivity, 80
signals, 26, 74, 75, 97
skeleton, 69, 76, 83
sodium, 34, 37, 50
solution, 19, 29, 34, 35, 36, 37, 38, 39, 40, 46, 107
solvents, 3, 4, 5, 6, 20, 34, 42, 53
sowing, 59, 60, 63, 65
spectroscopy, 3, 25, 30
stable complex, 106
structure, vii, viii, ix, 1, 3, 4, 15, 21, 26, 30, 34, 74, 91, 97, 99, 101, 102, 104, 108
substrates, 70, 73, 77, 81, 83, 123
sulfate, 8, 37, 46, 56
sulfur, viii, 37, 42, 68, 69
sulfuric acid, 5
symmetrical acetylenes, ix, 92, 93, 98, 99, 100, 102, 105, 106, 107, 109, 110, 111, 114, 123
synthesis, vii, ix, 6, 30, 34, 36, 37, 41, 42, 47, 54, 57, 61, 63, 64, 70, 74, 77, 78, 81, 83, 91, 92, 93, 103, 110, 114, 116, 118, 119, 122, 123, 124, 125, 126
synthetic analogues, viii, 68, 69

T

target, 6, 12, 35, 36, 39, 73, 80, 94, 96
temperature, 3, 6, 12, 14, 95, 106, 112
tetrahydrofuran, 5, 95, 96, 125

tetrasubstituted furans, vii, ix, 91, 92, 93, 96, 98, 99, 100, 102, 109, 110, 112, 113, 114, 117, 119, 125

transformations, vii, viii, 3, 6, 12, 13, 15, 16, 19, 20, 25, 26, 27, 28, 29, 39, 40, 49, 50, 58, 59, 61, 62, 64, 68, 69, 105, 108

transition metal, 69, 123

U

unsymmetrical acetylenes, vii, ix, 91, 92, 93, 110, 111, 112, 114

V

vanadium, 2, 6, 7, 9, 10, 11, 12, 13, 27, 32, 36, 39, 40, 45, 46, 47, 48, 57, 61

W

water, viii, 2, 5, 7, 8, 10, 16, 17, 20, 21, 22, 27, 30, 31, 32, 33, 34, 37, 38, 39, 40, 41, 42, 48, 60, 63, 65, 80

Y

yield, ix, 2, 5, 8, 9, 11, 12, 15, 18, 20, 21, 26, 27, 29, 31, 32, 34, 36, 37, 38, 39, 40, 41, 42, 71, 72, 73, 74, 79, 80, 81, 91, 94, 95, 96, 97, 98, 99, 101, 102, 105, 106, 109, 111, 112, 113

ω

ω-dicarboxylic acid esters, ix, 92

Hydrogen Peroxide: Detection, Applications and Health Implications

EDITORS: Gilberto Aguilar and Raphael A. Guzman

SERIES: Biochemistry Research Trends

BOOK DESCRIPTION: Hydrogen peroxide (H2O2) is a highly reactive compound produced by cells after secondary reactions as a result of the leakage of electrons from the electron transport chain in mitochondria or as a consequence of other enzymatic reactions. This book examines the detection, applications and health implications of hydrogen peroxide.

HARDCOVER ISBN: 978-1-62257-414-8
RETAIL PRICE: $209

Handbook of Polycyclic Aromatic Hydrocarbons: Chemistry, Occurrence and Health Issues

EDITORS: Guilherme C. Bandeira and Henrique E. Meneses

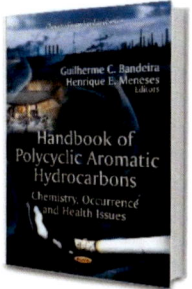

SERIES: Chemistry Research and Applications

BOOK DESCRIPTION: Polycyclic aromatic hydrocarbons (PAHs) are high molecular weight, aromatic compounds containing two or more benzene rings joined together in different ways. They belong to a group of persistent organic pollutants (POPs); are resistant to degradation; and can remain in the environment for long periods with the potential to cause adverse environmental and health effects.

HARDCOVER ISBN: 978-1-62257-473-5
RETAIL PRICE: $250